Mechano-chemical synthesis and characterisation of complex hydrides for solid state hydrogen storage

DISSERTATION

zur Erlangung des akademischen Grades

Doktoringenieur

(Dr.-Ing.)

vorgelegt

der Fakultät Maschinenwesen

der Technischen Universität Dresden

von

Dipl.Ing. Carine Rongeat

geb. am 28.05.1978 in Mâcon, Frankreich

Dresden 2011

Bibliografische Information der Deutschen Nationalbibliothek
Die Deutsche Nationalbibliothek verzeichnet diese Publikation in der Deutschen Nationalbibliografie; detaillierte bibliografische Daten sind im Internet über http://dnb.d-nb.de abrufbar.
1. Aufl. - Göttingen : Cuvillier, 2011
Zugl.: (TU) Dresden, Univ., Diss., 2011

978-3-86955-663-5

Nonnenstieg 8, 37075 Göttingen
Telefon: 0551-54724-0
Telefax: 0551-54724-21
www.cuvillier.de

1. Auflage, 2011
Gedruckt auf säurefreiem Papier

978-3-86955-663-5

Gutachter 1. Prof. Ludwig Schultz, Technische Universität Dresden, Deutschland

Gutachter 2: Prof. Asuncion Fernandez Camacho, Universidad de Sevilla, Spanien

Tag der Einrichtung: 6. Oktober 2010

Tag der Verteidigung: 13. Januar 2011

Abstract

Hydrogen storage technologies are gaining interest for the development of new sustainable energy carriers instead of fossil fuel and for reducing the release of greenhouse gases. Hydrogen is indeed a clean energy carrier but a safe and reliable storage is necessary. Solid state hydrogen storage in hydrides offers a good alternative for an efficient storage. Some intermetallic alloys can reversibly store hydrogen at room temperature but the gravimetric hydrogen density is low (ca. 2wt% H_2). In particular for mobile applications, a high gravimetric and volumetric storage capacity is necessary. These requirements have triggered a lot of interest for complex hydrides (metal alanates $M(AlH_4)_x$ or borohydrides $M(BH_4)_x$).

Here, reactive ball milling under hydrogen pressure is developed for the synthesis of complex hydrides from their decomposition products. Recording the pressure and temperature inside the vial during milling is used to analyse the synthesis reaction and the effect of different additives/dopants. Two different kind of complex hydrides were selected: $NaAlH_4$ as a model system and $Ca(BH_4)_2$ for its high hydrogen capacity.

Sodium alanate $NaAlH_4$ is a prototypical complex hydride that can store practically ca. 4wt% of hydrogen in the 100-150°C temperature range that is interesting for the operation of a PEM (Proton Exchange Membrane) fuel cell. This can be achieved only when adding a dopant to $NaAlH_4$. Although this compound is studied for several years, there is still no consensus on the doping mechanism. In this work, different metal and metal chloride dopants for $NaAlH_4$ are studied. The comparison of these dopants is used to better understand the catalytic and reaction mechanisms. The calalytic mechanism is different for hydrogen desorption and absorption. For desorption, $TiCl_3$ is found to be the best dopant due to the creation of a high number of vacancies formed by the decomposition of $TiCl_3$ during milling. These vacancies promote the diffusion processes inside the material. The formation of Al_3Ti may also act as a grain refiner preventing the grain growth during cycling and maintaining the nanocrystalline structure. For absorption, the best dopant is $CeCl_3$. The highest reaction rates obtained are related to the formation of the Al_4Ce phase that favours the absorption of hydrogen inside the material through the formation of Al-Ce-H complexes.

Nevertheless, despite its good reversibility, $NaAlH_4$ cannot be considered for mobile application because of its limited hydrogen capacity. Metal borohydrides have therefore been considered more recently due to their high gravimetric and volumetric stored hydrogen densities. Among them, $Ca(BH_4)_2$ (11.5wt% H_2) is interesting because of the predicted suitable thermodynamic properties, that would enable the release of hydrogen within the operation range of PEM fuel cells. A new synthesis route is investigated, using reactive ball milling of CaH_2 and CaB_6 under high hydrogen pressure (max 150 bar). The yield of formation of $Ca(BH_4)_2$ after milling is low (20-30%) and the combination of different techniques, in particular infrared spectroscopy, is necessary to identify $Ca(BH_4)_2$ in the milled samples. Nevertheless, the reversible decomposition/formation of $Ca(BH_4)_2$ is observed (60% yield) during further cycling and is obtained using TiF_3 as additive. The reversible H_2 sorption cannot be achieved with $TiCl_3$ as additive because of the formation of other products. The understanding of the difference observed when using these two additives is of crucial importance to further improve the kinetics of H_2 sorption and the use of $Ca(BH_4)_2$ for hydrogen storage.

Zusammenfassung

Die Wasserstoffspeicherung ist ein Eckpfeiler für eine moderne Energiewirtschaft, die durch eine geographische, saisonale und tageszeitliche Diskontinuität der erneuerbaren Energien, geprägt ist. Aufgrund der reichlichen Verfügbarkeit und des hohen Energieinhaltes von Wasserstoff wird er als attraktive Alternative zu fossilen Brennstoffen in Betracht gezogen. Wasserstoff ist ein sauberer Energieträger, sofern durch erneuerbare Energien gewonnen, durch dessen Einsatz die Treibhausgasemissionen deutlich gesenkt werden könnten. Jedoch müssen für eine wirtschaftliche und effiziente Nutzung sichere und verlässliche Speichermöglichkeiten entwickelt werden. Die Wasserstoffspeicherung im Festkörper wie beispielsweise in Metallhydriden stellt die effizienteste Speichermöglichkeit gegenüber der Speicherung in Druckgasflaschen oder in flüssiger Form dar. Einige intermetallische Verbindungen sind in der Lage Wasserstoff reversibel bei Raumtemperatur zu speichern. Deren gravimetrische Wasserstoffdichte ist mit max. 2 Gew.% jedoch zu gering. Insbesondere für mobile Anwendungen wird eine hohe gravimetrische als auch volumetrische Wasserstoffdichte gefordert. Aus diesem Grund stehen die komplexen Hydride (Alanate $M(AlH_4)_x$ und Borhydride $M(BH_4)_x$) seit einigen Jahren in besonderem Fokus der Materialforschung.

Für die Synthese dieser komplexen Hydride aus den Zersetzungsprodukten wurde in dieser Arbeit das Reaktivmahlen unter Wasserstoffdruck entwickelt. Mit Hilfe zweier Sensoren im Deckel des Mahlbechers konnten Druck und Temperatur während des Mahlprozesses in-situ gemessen werden. Anhand dieser Daten kann die Synthesereaktion überwacht und der Effekt von Additiven schon während der Herstellung studiert werden. Zwei verschiedene komplexe Hydride wurden für diese Arbeit aufgrund ihrer herausragenden Eigenschaften ausgewählt. Zum einen $NaAlH_4$ als Modellmaterial komplexer Hydride und zum andere $Ca(BH_4)_2$ aufgrund seiner extrem hohe gravimetrische Wasserstoffspeicherdichte.

Natriumalanat ($NaAlH_4$) ist der bekannteste Vertreter der komplexen Hydride, da es unter vergleichsweise moderaten Bedingungen ($T = 100$ bis 150°C, $p = 100$ bar) praktisch 4 Gew.% Wasserstoff reversibel im Temperaturbereich von 100 bis 150°C speichern kann. Durch den Zusatz eines Additivs wie $TiCl_3$ funktioniert das Material in einem Temperaturbereich, der für den Einsatz mit PEM Brennstoffzellen kompatibel ist. Obwohl das Natriumalanat schon seit einigen Jahren untersucht wird, ist der Mechanismus, den das Additiv bewirkt, noch immer nicht aufgeklärt. In dieser Arbeit wird daher der Einfluss unterschiedlichster Zusätze (Metalle und Metallchloride) auf die Eigenschaften des

Natriumalanats untersucht. Dazu wurden die unterschiedlichen Effekte der einzelnen Additive miteinander verglichen, um Rückschlüsse auf deren Einfluss zu ziehen. Eine wichtige Erkenntnis ist die Beobachtung, dass sich die Mechanismen für die Wasserstoffabsorption und -desorption unterscheiden. $TiCl_3$ zeigt den besten Effekt für die Wasserstoffdesorption. Aufgrund zahlreicher Leerstellen, die durch die Zersetzung des $TiCl_3$ während des Mahlprozesses entstehen, werden die Diffusionsprozesse im Material begünstigt. Das dabei gebildete Al_3Ti scheint zur Kornfeinung beizutragen, da das Kornwachstum während der Zyklen deutlich geringer ist, wodurch die Nanostruktur des Pulvers erhalten bleibt. Hingegen hat sich $CeCl_3$ als bester Zusatz zur Verbesserung der Bedingungen bei der Wasserstoffabsorption gezeigt. Bei Zugabe von $CeCl_3$ wird die höchste Reaktionsrate beobachtet, die auf die Bildung von Al_4Ce zurückzuführen ist, da diese Phase die Hydrierung im Material fördert.

Trotz der guten Reversibilität kann das $NaAlH_4$ aufgrund seiner begrenzten Wasserstoffkapazität nicht für mobile Anwendungen in Erwägung gezogen werden. In den vergangenen Jahren wurde intensive Forschung im Bereich der Borhydride durchgeführt, da diese sich durch eine extrem hohe gravimetrische und volumetrische Wasserstoffdichte auszeichnen. Aufgrund seiner berechneten thermodynamischen Eigenschaften, die die Operation mit PEM Brennstoffzellen ermöglichen würden, ist das Kalziumborhydrid ($Ca(BH_4)_2$, 11.5 Gew.% H_2) ein vielversprechender Kandidat zur Wasserstoffspeicherung im Festkörper. Mit dem Reaktivmahlen von CaH_2 und CaB_6 unter Wasserstoffdruck (max. 150 bar) wurde ein neuer Syntheseweg des $Ca(BH_4)_2$ untersucht. Die Ausbeute der Synthese ist mit etwa 20-30% relativ gering. Das hergestellte Pulver wurde mittels verschiedenster Techniken charakterisiert, wobei die Identifizierung des $Ca(BH_4)_2$ zweifelsfrei nur mit Infrarotspektroskopie gelang. Nichtsdestotrotz wurde die reversible Bildung bzw. Zersetzung des $Ca(BH_4)_2$ mit einer Ausbeute von etwa 60% durch die Zugabe von TiF_3 erreicht. Die reversible Sorption von Wasserstoff konnte mit $TiCl_3$ nicht erreicht werden, da hierbei unerwünschte Produkte gebildet wurden. Das Verständnis über den unterschiedlichen Einfluss der Additive ist von höchster Wichtigkeit, um die Kinetik der Wasserstoffaufnahme und –abgabe im $Ca(BH_4)_2$ weiter zu verbessern und somit einen potentiellen Einsatz in der Anwendung zu ermöglichen.

Content

Abstract ... *ii*
Zusammenfassung ... *iv*
Content ... *vi*
List of symbols and abbreviations ... *viii*

1 Introduction *1*

1.1 Context **2**
1.1.1 Hydrogen economy 2
1.1.2 Hydrogen storage 4
1.1.2.1. Physical storage 6
1.1.2.2. Chemical storage 7

1.2 The complex hydrides **9**
1.2.1 Sodium alanate, the model for complex hydrides 11
1.2.1.1. General properties 11
1.2.1.2. Thermodynamic properties 12
1.2.1.3. Reaction paths 13
1.2.1.4. Synthesis 14
1.2.1.5. Addition of dopants 15
1.2.1.6. Catalytic mechanism 17
1.2.2 $Ca(BH_4)_2$: a high capacity storage material 21
1.2.2.1. Hydrogen sorption properties of $Ca(BH_4)_2$ 22
1.2.2.2. Crystal structure 23
1.2.2.3. Synthesis 24
1.2.2.4. Reaction paths and effect of dopants. 24
1.2.3 Open questions in complex hydrides and aim of this work 27

2 Experimental *29*

2.1 High pressure reactive ball milling **29**

2.2 Structural characterisation **31**
2.2.1 X-ray Diffraction (XRD) 31
2.2.2 Microscopy 32
2.2.3 X-ray Photoelectron Spectroscopy (XPS) 34
2.2.4 Vibrational spectroscopy 35
2.2.5 BET (Brunauer-Emmett-Teller) analysis 36

2.3 Thermal analysis **37**
2.3.1 Differential scanning calorimetry (DSC) 37
2.3.2 Thermogravimetry (TG) 40
2.3.3 Pressure-composition Isotherms (PCI) 41
2.3.4 Thermal Desorption Spectroscopy (TDS) with mass spectrometry 43

3 $NaAlH_4$: a model system *45*

3.1 Synthesis by reactive ball milling **46**
3.1.1 Influence of the milling parameters on the synthesis process 46
3.1.2 Influence of the hydrogen pressure 50
3.1.3 Influence of the dopant 53

3.1.3.1. Effect of dopant content 53
3.1.3.2. Study of different dopants 55
3.1.4 Conclusions on synthesis by reactive ball milling 58

3.2 Thermodynamic and kinetic properties of doped-$NaAlH_4$: comparison of dopants 58
3.2.1 Decomposition process 58
3.2.2 Thermodynamic properties 61
3.2.3 Kinetic properties 66
3.2.3.1. Decomposition 66
3.2.3.2. Re-absorption and cycling 69
3.2.4 Summary of the dopant effect on thermodynamic and kinetic properties 72

3.3 Catalytic mechanism 73
3.3.1 Study of the catalysis of H_2 desorption 73
3.3.1.1. Analysis of decomposition products 73
3.3.1.2. Morphology of decomposed samples 74
3.3.1.3. XPS analysis 77
3.3.2 Catalysis for the H_2 absorption 80
3.3.3 Proposed catalytic mechanism 81

4 New high capacity complex hydride: $Ca(BH_4)_2$ 83

4.1 Synthesis by reactive ball milling 84
4.1.1 Milling process 84
4.1.2 Identification of the phases formed during milling 88
4.1.2.1. In-situ XRD 88
4.1.2.2. Vibrational spectroscopy 89
4.1.2.3. TEM and EELS 91
4.1.3 Conclusions on synthesis 94

4.2 Decomposition of $Ca(BH_4)_2$: kinetics and mechanism. 94
4.2.1 Study of pure $Ca(BH_4)_2$ 94
4.2.2 Decomposition of as-milled sample 96
4.2.2.1. DSC 96
4.2.2.2. TDS 97
4.2.2.3. XRD analysis of the decomposition route 99
4.2.3 Summary of the results on $Ca(BH_4)_2$ decomposition 100

4.3 Effect of additives on the reversibility 101
4.3.1 Re-hydrogenation of decomposed samples in Sieverts' and DSC apparatus 101
4.3.2 Analysis of the phases formed after re-hydrogenation 103
4.3.3 Discussion on the higher efficiency of TiF_3 for re-hydrogenation 106
4.3.4 Conclusions on the H_2 sorption properties of $Ca(BH_4)_2$ 107

5 Summary and outlook 109

6 References 115

7 Annexes 133
List of publications.....145
List of figures.....147
List of tables.....149
Acknowledgements.....151

List of symbols and abbreviations

ΔG	free energy of a reaction
ΔH	enthalpy of a reaction
$\Delta_f H^0$	enthalpy of formation of a compound
Δn	variation of the number of moles H_2 during milling
Δp	pressure drop observed during milling
ΔQ	heat exchanged during a reaction
ΔS	entropy of a reaction
θ	diffraction angle
BET	Brunauer-Emmet-Teller analysis
BPR	ball-to-powder ratio
DFT	Density Functional Theory
DOE	US Department of Energy
DSC	Differential Scanning Calorimetry
E^a	activation energy
EDS/EDX	Energy Dispersive Spectroscopy (X-rays)
EELS	Electron Energy Loss Spectroscopy
HP-DSC	High-Pressure Differential Scanning Calorimetry
ICSD	Inorganic Crystal Structure Database
IR	infrared
k	rate constant
MH	Metal Hydride
MOF	Metal Organic Framework
MS	Mass Spectrometer
NMR	Nuclear Magnetic Resonance
p	pressure
PCI	Pressure-Composition Isotherms
pcT	pressure-composition-temperature diagram (see PCI)
PEMFC	Proton Exchange Membrane Fuel Cell
Q	wave vector
R	ideal gas constant
RBM	Reactive Ball Milling

RHC	Reactive Hydride Composite
SE	Secondary Electrons
SEM	Scanning Electron Microscopy
T	temperature
T_{dec}	decomposition temperature of a hydride
T_{eq}	equilibrium temperature in fixed conditions
T_{ig}	ignition temperature
T^{m}	melting point
T^{struc}	temperature for structural transition
TDS	Thermal Desorption Spectroscopy
TEM	Transmission Electron Microscopy
TG	Thermogravimetry
TM	Transition Metal
XANES	X-ray Absorption Near Edge Spectroscopy
XAS	X-ray Absorption Spectroscopy
XPS	X-ray Photoelectron Spectroscopy
XRD	X-ray Diffraction

1. Introduction

Towards Hydrogen Economy: the storage challenge

The recent awareness of the climate change and concern about the fossil fuel availability have underlined the necessity of developing new energy concepts to face the future energy supply challenges. A new economy can be developed based on hydrogen as a clean energy carrier but implies a complete change of consumption habits compared to an oil-based society. Oil, coal or gas are primary energy sources, because they are directly available without conversion in nature. Hydrogen, as electricity, is a secondary energy source also called energy carrier. This means that it cannot directly be extracted from earth. It is obtained by conversion of a primary energy source and can be used to transport and store more efficiently energy than the primary sources, e.g. solar or wind energy. Hydrogen is an abundant element and H_2 has the highest energy-to-weight ratio, its combustion releases only water and it can be produced at any place of the world. Hydrogen offers a good partnership with electricity; the latter being a well established energy carrier but difficult to store efficiently. Conversion to hydrogen is then a good alternative for energy storage and appears to be particularly interesting for transportation over long distances where batteries cannot be used.

1.1 Context

1.1.1 Hydrogen economy

The hydrogen cycle is given in **Figure 1.1** and shows the different steps necessary for the use of hydrogen as an energy carrier [1, 2]. Each step of this cycle, such as hydrogen production, transport and storage, and combustion, has to be newly developed and established worldwide to finally replace the oil-based economy.

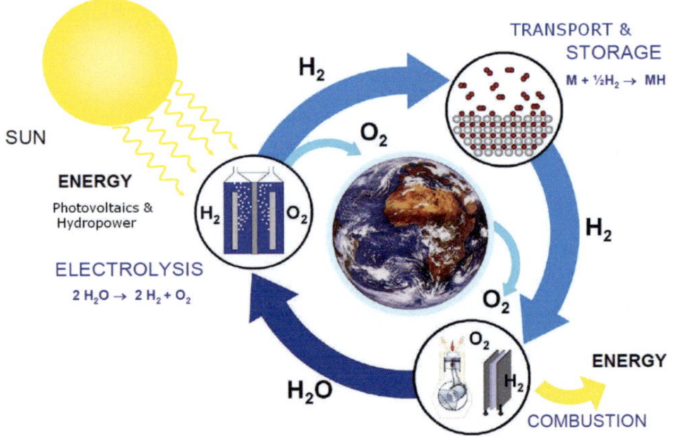

Figure 1.1: The hydrogen cycle (from [1]).

The production of hydrogen is the first challenge, because, despite its high abundance, because hydrogen is mainly present bonded to other elements (mostly oxygen), e.g. in water (H_2O). Hydrogen can be produced easily by reforming hydrocarbons but this process relies also on fossil fuels (oil). To ensure a clean process, hydrogen can be produced from the electrolysis of water using renewable energies sources (like wind or solar) for the supply of electricity. The electrolysis of water is a well known technology and consists of splitting water molecules to oxygen and hydrogen gas; it produces high purity hydrogen. However, the process requires a high quantity of energy (minimum of 39.4 kWh kg^{-1} H_2 [1]). There are then important concerns about the supply of electricity. Nowadays, the electrolysers have already a high efficiency (82%); this means that large technical challenges have to be overcome to reduce further the energy losses.

Energy can be liberated from hydrogen by direct combustion (heat). On the other hand, hydrogen can also be used to produce electricity using a fuel cell. This last solution is preferred for mobile applications. The fuel cell stacks are coupled to an electrical motor to power vehicles.

Some examples of fuel cell powered vehicles have been already developed and constructed by the automotive industry. The principle of a fuel cell is known since 1839 [3] and consists in the reverse reaction of water electrolysis. Oxygen and hydrogen react together to produce electricity and water. The most commonly used fuel cell is the proton exchange membrane fuel cell (PEMFC) (**Figure 1.2**) [4]. Nevertheless, there are still technical issues that limit the development of such technology. There is still space for improvement in fuel cell technology, for example by the use of non-noble catalysts [5]. With the present fuel cell technology, 4 kg of hydrogen are needed to power a car with an autonomy range of 400 km.

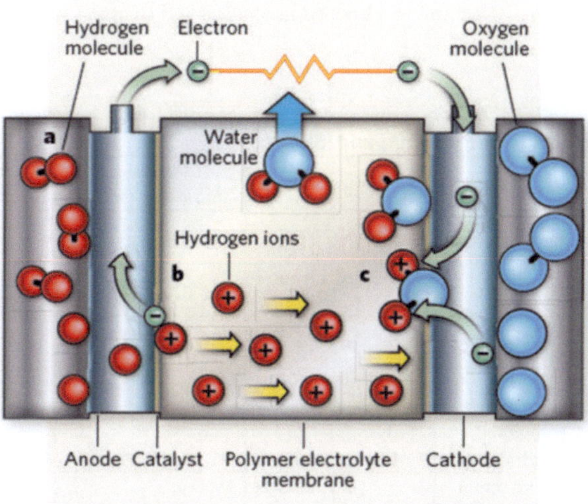

Figure 1.2: Principle of a PEM fuel cell (from [6]).
a: Hydrogen is channelled to the anode. b: a catalyst lining the anode causes the hydrogen atoms to split into hydrogen ions (H^+) and electrons. A polymer electrolyte membrane allows the hydrogen ions to pass between the anode and the cathode, but the electrons must pass through an external circuit to reach the cathode, creating a current. c: oxygen is channelled to the cathode, where it reacts with electrons and hydrogen ions to form water as the only side product.

Hydrogen production and combustion can be greatly improved but the bottleneck for the hydrogen cycle is transportation and storage [7]. As hydrogen is a gas at normal temperature and pressure, its storage and transportation is less convenient than liquid gasoline. Transportation can be done in pipelines but is usually limited to short distance. This means that efficient storage methods have to be found. Under standard conditions (room temperature and atmospheric pressure), 4 kg of hydrogen represent a volume of 45 m^3. Different methods have been proposed to reduce this volume. The largest challenge is for mobile applications since the space in a car is very limited. The different storage technologies, already available or in development, are described in the next section.

1.1.2 Hydrogen storage

After this short introduction to the hydrogen economy, it appears clearly that storage is a major issue. As already mentioned, hydrogen exists in the gas phase at standard conditions, which call for the development of efficient storage methods. There are different technologies available for the storage of hydrogen. They can be classified in physical and chemical storage [1, 7, 8] (**Table 1.1** and **Figure 1.3**).

Table 1.1: Classification and properties of different hydrogen storage technologies (adapted from [4] and [8])

Storage type	*Name*	*Examples*	*Maximum H_2 capacity (wt%)*	*Operating temperature (°C)*
Physical storage	Liquid hydrogen	-	100	-253
	Compressed gas	350 to 700 bar	100	25
	Cryo-adsorption	activated carbon MOF [a]	6.5 7.5 [b]	-200
Chemical storage	Chemical hydrides	$NaBH_4$	7.3	25
	Liquid hydrocarbons	Cyclohexane Decalin	7.1 7.2	300
	Interstitial hydride	$LaNiH_x$, $FeTiH_x$, Laves phases	2	0-30
	Salt-like metal hydride	MgH_2	7.6	330
	Complex hydrides	$NaAlH_4$ $LiBH_4$	5.5 18.4	70-170 400

[a] MOF: Metal Organic Framework
[b] for MOF177 with 5900 m^2 g^{-1} surface area

The US department of energy (DOE) has enumerated system requirements for the hydrogen storage media (**Table 1.2**) [9]. These targets are commonly used as guideline in the hydrogen storage research area, although very debated. The values have been set to resemble the use of a conventional gasoline vehicle (e.g. 5 min refuelling time), and to fit the technical requirement of a PEM fuel cell, which are thought to be used in cars (e.g. operating temperature). One important criterion is the gravimetric storage density: the target is 5.5 wt% H_2 for the tank system in 2015.

Table 1.2: DOE targets (revised) for on-board hydrogen storage systems of light-duty vehicles (Feb. 09) [9]

Storage parameters	*Units*	*2010*	*2015*	*Ultimate*
System gravimetric capacity: usable, specific-energy from H_2 *(net useful energy/max system mass)*	kWh kg^{-1} *(kg H_2 kg^{-1} system)*	1.5 *(0.045)*	1.8 *(0.055)*	2.5 *(0.075)*
System volumetric capacity: usable energy density from H_2 *(net useful energy/ max system volume)*	kWh L^{-1} *(kg H_2 L^{-1} system)*	0.9 *(0.028)*	1.3 *(0.040)*	2.3 *(0.070)*
Storage system cost & fuel cost	\$ kWh^{-1} net (\$ kg^{-1} H_2) \$ gge^{-1} at pump[a]	4 (133) 2-3	2 (67) 2-3	? - 2-3
Durability/operability				
• Operating ambient temperature	°C	-30/50	-40/60	-40/60
• Min/max delivery temperature	°C	-40/85	-40/85	-40/85
• Cycle life (1/4 tank to full)	Cycles	1000	1500	1500
• Minimum delivery pressure from storage system (FC)	atm	4	3	3
• Max delivery pressure from storage system	atm	100	100	100
Charging/discharging rate				
• System fill time (for 5 kg H_2)	min *(kg H_2 min^{-1})*	4.2 *(1.2)*	3.3 *(1.5)*	2.5 *(2)*
• Minimum full flow rate	(g s^{-1}) kW^{-1}	0.02	0.02	0.02
• Start time to full flow (20°C)	s	5	5	5
• Start time to full flow (-20°C)	s	15	15	15
• Transient response 10%-90% and 90%-0%	s	0.75	0.75	0.75
Fuel purity (H_2 from storage)	% H_2	99.99 (dry basis)		
Environmental health and safety				
• Permeation & leakage	Scc h^{-1} [b]	meets or exceeds applicable standards		
• Toxicity	-			
• Safety	-			
• Loss of usable H_2	(g h^{-1}) kg^{-1} H_2 stored	0.1	0.05	0.05

[a] gge: gallon gasoline equivalent (amount of alternative fuel to equal the energy content of 1 gallon gasoline)

[b] Scc h^{-1}: standard cubic centimetre per hour (gas flow unit in standard conditions of pressure, temperature and humidity)

The physical storage can be divided in three major technologies: the storage in compressed H_2 gas, the storage of liquid H_2 (cryotanks) and the adsorption of H_2 on materials with high surface area (cryo-adsorption). The chemical storage contains mainly the solid state storage in hydride compounds (metal or complex hydrides) or in hydrocarbons (reforming). The first two physical storage methods are the most established technologies, very similar to gasoline technology, including a tank that can be refilled on-board. The chemical storage methods include reversible and irreversible reactions. The irreversible reactions imply that the tank has to be refilled externally. The reversible reactions allow refuelling on-board as for gas or liquid hydrogen. A summary of the state-of-the-art performances of different storage methods is given in **Figure 1.3**. For compressed or liquid hydrogen, the maximum gravimetric or volumetric capacities have been reached, although technical improvements in the system might be possible. For the hydrides, it is expected that the H_2 capacity can be still greatly improved by the development of new compounds that are described in more detail below.

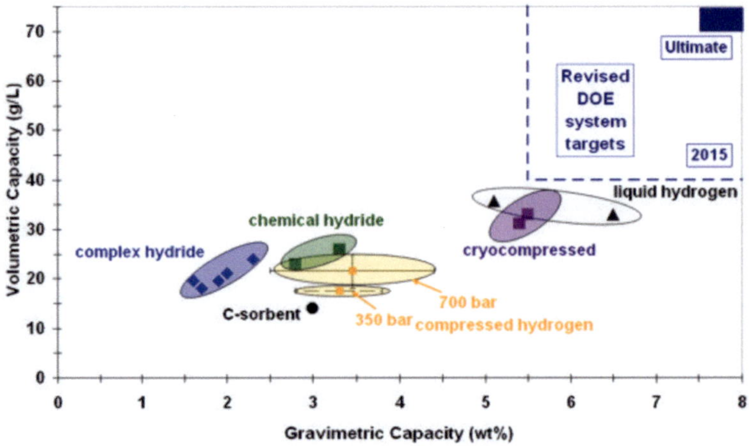

Figure 1.3: State-of-the-art of hydrogen storage technologies (values given for storage systems) [9]

1.1.2.1. Physical storage

To store 4 kg of hydrogen in the gaseous state in a sufficiently small volume, it is necessary to use high pressures (350-700 bar). The use of lower values would not be relevant because of too small amounts of hydrogen stored. The use of higher values would require too much energy compared to the additional quantity of hydrogen stored, because of the deviation from ideal gas behaviour. The compressed H_2 tank is the most established technology, which has been used with 300 bar tanks for a long time for the transportation and storage of H_2. Nevertheless, the development of new tank designs is necessary to resist the high pressures required for mobile applications (350-700 bar). This is achieved using high-strength materials like carbon composites. A 700 bar tank made of these materials weighs only 135 kg (for 4.2 kg H_2 stored) while an equivalent steel tank would weigh 700 kg [8]. This storage method has the best overall performance at present and also the highest maturity. However, it still requires a high amount of energy to compress the gas and technical issues are faced for the integration in cars. There are also questions about the public acceptance for using such high pressure tanks in cars and the cost of a reasonable safety system.

Liquid hydrogen has also been considered for storage because it has several advantages like easy refuelling and high mass density, in particular compared to compressed gas tanks. However, the liquid region of hydrogen in the phase diagram is small [10] and cryo-tanks have to keep a temperature of -253°C (21 K). Such cryo-tanks have been developed already. The main disadvantages of this technique are the significant gas losses and the high energy consumption for liquefying hydrogen. The losses are related to the boil-off effect and the evaporation of hydrogen gas from the tank. This has to be taken into consideration for long time storage with the necessity of venting hydrogen to prevent a pressure increase in the tank. In addition, a very efficient insulation is required to keep the very low temperature. All these drawbacks make liquid hydrogen not attractive for mobile applications.

The last method for physical storage is the adsorption of H_2 on the surface of large surface-area materials [8, 11]. This type of storage is classified with the physical method because it involves the storage of molecular hydrogen that has only weak interactions with the adsorbent. This method requires low temperature, -196°C (77 K), and/or high pressures. In general, the capacity of hydrogen adsorbed depends on the specific surface area and the pore structure and size of the adsorbents. Different classes of adsorbents have been developed: carbon materials (activated carbon, graphite), metal-organic frameworks (MOF), zeolites. MOF materials appear to have the highest gravimetric storage density with 7.5wt% H_2 reported for MOF177 (surface area 5900 $m^2 g^{-1}$) at -196°C [8]. However, this performance cannot be obtained at room temperature and a high quantity of energy is necessary for the adsorption implying the necessity of a large quantity of liquid nitrogen. This causes severe engineering challenges and adds high costs that make this solution not attractive for mobile applications.

1.1.2.2. Chemical storage

In contrast to physical storage methods, the chemical methods involve the storage of atomic hydrogen and the breaking and formation of chemical bonds.

To ease handling, the use of liquid hydrogen carriers like hydrocarbons has been proposed [8, 12]. This includes for example methanol or cyclic hydrocarbons like benzene or cyclohexane. The cyclic hydrocarbons may be used directly in tanks but the release of hydrogen requires too high temperature to be used in mobile applications. Another route is the production of hydrogen on-board from the reforming of liquid hydrocarbons. Obviously, the addition of a reforming system in a car presents a technical issue. In addition, CO gas may be released, which is very poisonous for the fuel cell. Easier is then the use of methanol, of which reforming is less difficult than for other hydrocarbons. The use of methanol in a direct methanol fuel cell is also possible. This fuel cell has a lower power density than the hydrogen fuel cell but the technology is already available and can be an alternative for mobile applications.

Even more attractive is the storage of hydrogen in the solid state, i.e. in hydride compounds. The use of solids avoids the use of high pressures or low temperatures and makes it a reliable and safe storage method. The target of 5.5 wt% H_2 for the tank system means that the hydride compound must contain more than 8 or even 10 wt% H_2. Different classes of hydrides can be used. The so-called chemical storage or hydrolytic storage involves the irreversible reaction of a hydride with water. For example, the reaction of $NaBH_4$ with water liberates 7.3wt% H_2 with the formation of NaOH and H_3BO_3. Nevertheless, onboard refuelling is not possible with this system. The use of reversible hydrides appears then as a more interesting technology, competing with the compressed gas method. To find suitable materials for hydrogen storage, the hydrides are evaluated from their thermodynamic and kinetic properties. The thermodynamic properties (e.g. enthalpy of hydride formation/decomposition) give the operating temperature and pressure. The kinetic properties determine the reaction rates.

The thermodynamic properties of hydrides can be described by a phase diagram (**Figure 1.4**). For a conventional metal hydride, the first part of the diagram for small hydrogen content shows an increase of the amount of hydrogen (solid solution α) with pressure when hydrogen enters in interstitial sites. Once the metal hydride phase (β) is formed, a plateau is observed corresponding to the two-phase region (α and β). Finally, when the metal is fully converted, the hydrogen capacity reaches its maximum value (β phase) and the pressure increases again. The pressure of the plateau at a certain temperature is called equilibrium pressure and depends on the thermodynamic properties of the hydride, i.e. the enthalpy ΔH and the entropy ΔS of hydriding / dehydriding. The entropy is quite similar for all materials because it mainly depends on the evolution of gaseous hydrogen ($\Delta S \sim 130$ kJ K^{-1} mol^{-1} H_2) [10]. The enthalpy of the reaction is then an additional key parameter, together with the gravimetric density, for the selection of a suitable hydrogen storage material. The enthalpy of the reaction should be in the 30-60 kJ mol^{-1} H_2 range for the release of hydrogen between 50 and 150°C [10, 13], as required for mobile applications. This range also ensures rehydrogenation under reasonable and technically achievable conditions (pressures below 100 bar H_2 in the operating temperature range). In addition, a low enthalpy of the reaction is favourable for tank application because it reduces the (waste) heat released during absorption. The heat management is one of the technical challenges for tank integration and mobile applications [8].

Hydrides can have very high hydrogen capacities, the hydrogen sorption reactions can be reversible and fast, and performed at suitable temperature for mobile application. Unfortunately at present, no hydride fulfils all these properties at the same time. The conventional metal hydrides ($LaNi_5H_6$, Ti-Fe-H, Ti-Cr-V-H and associated compounds called AB_2 or AB_5 alloys) are used for a very long time already (e.g. in Ni-MH batteries). The hydrogen sorption is very fast and the cycle life is very long. However, the gravimetric density is limited to roughly 2wt% H_2 that is much too low for mobile applications. By contrast, a very high capacity can be achieved by using MgH_2 (7.6wt% H_2). Magnesium has a lot of advantages: abundance, low cost, light weight. A massive effort in research allows now very fast hydrogen uptake/release [14-17]. This has been obtained by using nanostructured compounds (e.g. obtained by ball milling) and the addition of efficient catalysts like Nb_2O_5. The major drawback of MgH_2 is the very high enthalpy of reaction (78 kJ mol^{-1} H_2), which implies a high operating temperature (>300°C) in principle not compatible with a PEM fuel cell.

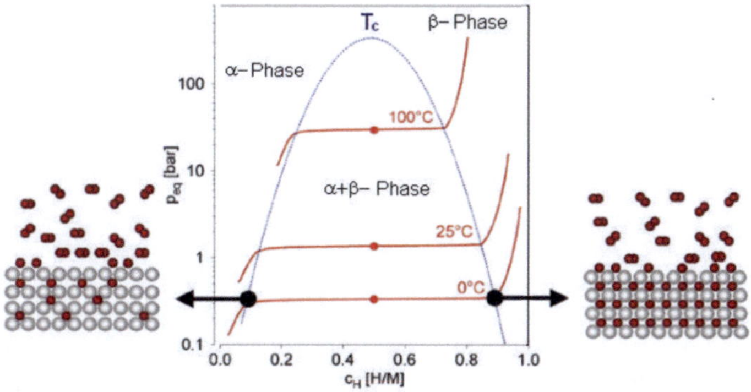

Figure 1.4: Phase diagram of a conventional metal hydride (adapted from [10]). α: solid solution H in metal and β: hydride phase.

Another class of hydride, the complex hydrides (or covalent hydrides) are considered for hydrogen storage since the work of Bogdanovic and Schwickardi [18] in 1997. These compounds are promising systems for solid state hydrogen storage in particular because of their high storage capacity, and are the topic of this work. They are presented in more detail in the next paragraph with a focus on the compounds studied here: $NaAlH_4$ and $Ca(BH_4)_2$.

1.2 The complex hydrides

The complex hydrides receive their names from the presence of a complex anion $[AH_4]^-$ which forms a tetrahedron with the metal in the middle and four hydrogen atoms on top [19]. They are described by the general formula $M(AH_4)_n$ where M is an alkaline or alkaline-earth element and A a metal from column III of the periodic table. To obtain the highest gravimetric density possible (**Table 1.2**), the complex hydrides considered are composed of light elements, i.e. M= Li, Na, K, Ca or Mg and A=Al or B. The $M(AlH_4)_n$ compounds are called alanates (metal tetrahydroaluminate) and the $M(BH_4)_n$ are called borohydride or boranates (metal tetrahydroboranate). Another class of hydrides is sometimes classified as complex hydride; this is the amide/imide compounds ($M(NH_2)_n/M_{2/n}(NH)_2$). The release of 6 wt% H_2 at 200°C reported for $LiNH_2$ [20] makes this compound very attractive. However, the evolution of NH_3 is often observed, limits the reaction reversibility and can poison the fuel cell. In the complex hydrides, H is strongly, covalently bonded to a metal atom. Breaking this bond means to completely decompose the compound into the elements according to one of the equations (1.1) to (1.4).

$$M(AlH_4)_n \rightarrow MH_n + nAl + 3n/2\, H_2 \quad \textbf{(1.1)}$$

$$M(BH_4)_n \rightarrow MH_n + nB + 3n/2\, H_2 \quad \textbf{(1.2)}$$

$$M(BH_4)_n \rightarrow MB_n + 2nH_2 \quad \textbf{(1.3)}$$

$$M(BH_4)_2 \rightarrow 2/3\, MH_2 + 1/3\, MB_6 + 10/3\, H_2 \quad \textbf{(1.4)}$$

These reactions can in reality hide multi-step processes with the formation of intermediate compounds. It can then be very difficult to re-form the hydrides despite their high stability. This is mainly related to the low reactivity of the decomposition products (e.g. amorphous boron in reaction (1.2)) and sluggish kinetics. Therefore, the complex hydrides have attracted interest for storage only after the demonstration that $NaAlH_4$ (sodium alanate) can reversibly release/uptake hydrogen through the overall reaction (1.1) at 150°C if $TiCl_3$ is used as additive [18]. Therefore, the complex hydrides are considered the more promising materials for hydrogen storage and are widely studied. $NaAlH_4$ is obviously the most studied and is used as model compound. More details about this hydride are given in section 1.2.1. Other types of alanates such as $LiAlH_4$ [21, 22], $Mg(AlH_4)_2$ [23, 24] or $Ca(AlH_4)_2$ [13, 25] have also been studied but have shown less interesting properties than $NaAlH_4$.

The metal borohydrides are very attractive materials with an even higher hydrogen capacity, in particular $LiBH_4$ (18.4 wt% H_2 [26]). However, they suffer from the difficulty of reversing the decomposition reaction and from slow kinetics. The properties and reaction mechanisms are more disparate than for the alanates as shown by the different decomposition routes observed (reaction (1.2) to (1.4)). For $LiBH_4$ or $NaBH_4$ (reaction (1.1)), improvements have been achieved for the kinetics [26-28] but they have a high reaction enthalpy (-74 KJ mol^{-1} for $LiBH_4$ [29] and -188.6 KJ mol^{-1} for $NaBH_4$ [30]). Then, the decomposition occurs at a too high temperature [26].

Therefore, a lot of work is now focused on $Ca(BH_4)_2$ (reaction (1.4)) [31] or $Mg(BH_4)_2$ (reaction (1.3)) [32-34], for which lower reaction enthalpies have been predicted. In particular, $Ca(BH_4)_2$ has several advantages that are described below (see §1.2.2). Nevertheless, there is still a lack of knowledge for the borohydrides, which is filled slowly by the on-going research work. For example, most of the metal borohydrides studied so far release hydrogen at quite high temperatures [19, 35] but it is not known whether the compounds are too stable (high enthalpy of dehydrogenation) or their decompositions are kinetically hindered. In addition, no dopant has been found up to now that would be as efficient as $TiCl_3$ for $NaAlH_4$.

Different concepts have also been reported to obtain a material with suitable kinetic and thermodynamic properties for storage applications. One main concept is the destabilisation of the borohydride, like $LiBH_4$, in order to decrease the reaction enthalpy. This is done by the addition of another hydride to form a so-called reactive hydride composite (RHC) that changes the decomposition products. The first composite reported is $LiBH_4$ + MgH_2 (reaction (1.5)) [36, 37].

$$LiBH_4 + 1/2\,MgH_2 \rightarrow LiH + 1/2\,MgB_2 + 2H_2 \qquad \textbf{(1.5)}$$

The decomposition of this composite leads to the formation of LiH and MgB_2. Reaction (1.5) has a lower enthalpy of reaction than pure $LiBH_4$ because of the formation of MgB_2; this explains why this composite is also called destabilised hydride. The formation of MgB_2 improves also the reversible absorption of hydrogen because B in MgB_2 is more reactive than B alone. The kinetic properties of this composite are dramatically improved by the addition of different compounds, e.g. Zr-based additives have been shown to be very efficient through the formation of ZrB_2 that favours the nucleation of MgB_2 during decomposition [38, 39]. This opens a new field of research aiming to tune the thermodynamic properties of the materials to fulfil the requirements of mobile applications [40]. Several composites have been explored by theoretical work to find reactions with suitable enthalpies [13, 41, 42]. Experimentally, $Ca(BH_4)_2$ + MgH_2 [43, 44] and $LiBH_4$ + YH_3 [45] are the two composite systems that have the most promising properties so far. Nevertheless, little is known about pure borohydride compounds other than $LiBH_4$ and $NaBH_4$. In particular for $Ca(BH_4)_2$, it is necessary to better understand the properties of the pure compound before applying the reactive hydride composite concept.

1.2.1 Sodium alanate, a model for complex hydrides

1.2.1.1. General properties

Sodium alanate $NaAlH_4$ is one of the most studied complex hydrides since Bogdanovic and Schwickardi [18] have demonstrated the possible reversible sorption of hydrogen using a Ti-based dopant. $NaAlH_4$ has a high hydrogen capacity with a maximum release of 7.5 wt% H_2. The decomposition reaction and the reverse absorption of hydrogen proceed in several steps according to the following equations:

$$3NaAlH_4 \leftrightarrow Na_3AlH_6 + 2Al + 3H_2 \quad (3.7\text{wt\% } H_2) \qquad \textbf{(1.6)}$$

$$Na_3AlH_6 \leftrightarrow 3NaH + Al + 3/2\,H_2 \quad (1.9\text{wt\% } H_2) \qquad \textbf{(1.7)}$$

$$NaH \leftrightarrow Na + 1/2\,H_2 \quad (1.9\text{wt\% } H_2) \qquad \textbf{(1.8)}$$

The last step of decomposition [reaction (1.8)] occurs at a too high temperature (~400°C) to be considered for practical use. For the two first steps, the theoretical amount of hydrogen released is 5.6 wt% H_2 and the maximum experimental capacity obtained is lower due to the necessary addition of a dopant. For $NaAlH_4$, the term "dopant" is preferred to "catalyst" because the compound reacts with $NaAlH_4$ (see below). The actual catalyst is not known as well as the real catalytic mechanism. In this section, any compound added to $NaAlH_4$ is denoted "dopant" and the mechanism of kinetic improvement is denoted "catalysis" or "catalytic mechanism".

$NaAlH_4$ has a tetragonal crystal structure with space group *I4₁/a* [46] and a melting point of about 180°C. Na_3AlH_6 displays two possible structures: at low temperature α-Na_3AlH_6 and at high temperature β-Na_3AlH_6. The α phase is monoclinic with space group $P2_1/n$ and the β phase is cubic with space group *Fm-3m* [47, 48]. The temperature for the phase transformation is *ca.* 250°C [49]. More information on the crystal structure is given in **Annexe A**.

1.2.1.2. Thermodynamic properties

Sodium alanate is the only complex hydride known at present to have favourable thermodynamic properties and reasonable gravimetric hydrogen capacity for using in mobile applications. The occurrence of different steps during decomposition increases the difficulty in understanding the reaction mechanisms. Bogdanovic and coworkers [18, 50] have obtained PCI (Pressure Composition Isotherms) curves giving the phase diagram and showing the two equilibrium plateaus corresponding to the two-step reaction (**Figure 1.5**). The plateau related to the second absorption step [reaction (1.7)] is at high pressure explaining the difficulty to re-hydrogenate the system even at high temperatures.

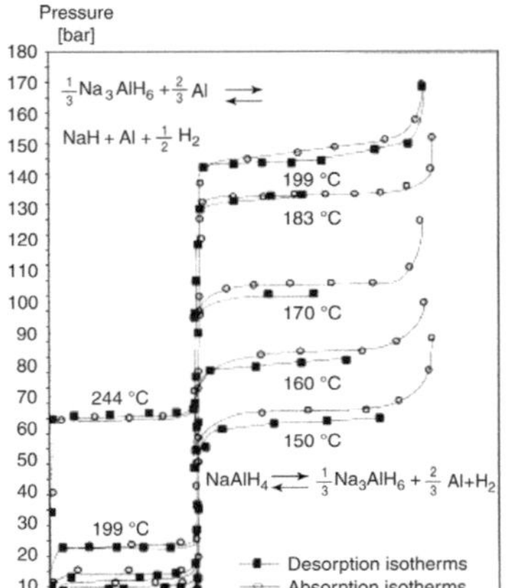

Figure 1.5: PCI curves for $NaAlH_4$ doped with 2 mol% $Ti(OBu)_4$ (Titanium (IV) butoxide) (from Bogdanovic et al. [50])

From these PCI and applying the van't Hoff equation (see chapter 2), the reaction enthalpy is determined to be 37 kJ mol^{-1} for the first step of the decomposition [eq. (1.6)] and 47 kJ mol^{-1} for the second step [eq. (1.7)]. These values are considered as reference values and imply decomposition temperatures for 1 bar H_2 at ~30°C and about 110°C for first and second decomposition steps, respectively [51]. Slightly different values for the enthalpy of reaction, in particular for equation (1.7), are given by different authors (**Table 1.3**). Some errors might be due to confusion in energy units per mol hydride or per mol H_2 [52]. The values reported in **Table 1.3** are given in kJ mol^{-1} (hydride) as written in the different publications. For reaction (1.6), the enthalpy is equal in kJ mol^{-1} or kJ mol^{-1} H_2 because one mol of H_2 is involved in the reaction. For reaction (1.7), a factor of 2/3 may be applied to convert from kJ mol^{-1} to kJ mol^{-1} H_2, likely to the highest values.

Table 1.3: Enthalpies of reaction for equation (1.6) and (1.7) reported in selected references.

Reaction	*Enthalpy of reaction (kJ mol^{-1})*	*Reference*
$NaAlH_4 \rightarrow 1/3\,Na_3AlH_6 + 2/3\,Al + H_2$	-37	[50]
	-36	[52]
	-45.5	[30]
	-31	[53]
	-34.5	[54]
	-27.8	[55]
	-36	[49]
$Na_3AlH_6 \rightarrow 3NaH + Al + 3/2\,H_2$	-47	[50]
	-69.6	[52]
	-89	[30]
	-56	[53]
	-71.4	[54]
	-67.7	[55]
	-41.4	[49]

Note: For reaction (1.6), the enthalpy is equal in kJ mol^{-1} or kJ mol^{-1} H_2 because one mol of H_2 is involved in the reaction. For reaction (1.7), a factor of 2/3 may be applied to convert from kJ mol^{-1} to kJ mol^{-1} H_2, likely to the highest values.

1.2.1.3. Reaction paths and kinetics

Differential Scanning Calorimetry (DSC) studies show the sequence of the different reaction steps during heating undoped $NaAlH_4$ under Ar atmosphere [49, 56] (**Figure 1.6a**). In the temperature range 160-280°C, four endothermic peaks occur. The first peak at ~180°C is related to the melting of $NaAlH_4$. The second peak (small) at 217°C is ascribed to the dissolution of Na_3AlH_6 in the molten $NaAlH_4$ [eq. (1.6)]. The third peak corresponds to the crystalline transformation of Na_3AlH_6 at 252°C. The fourth peak at 267°C is related to the decomposition of Na_3AlH_6 to NaH [eq. (1.7)]. Another peak around 400°C can be related to the decomposition of NaH to Na [eq. (1.8)]. This sequence has been confirmed by thermogravimetric analysis under Ar [57]. The two-step reaction has also been followed by in-situ X-ray diffraction (XRD) [58, 59]. The melting of $NaAlH_4$ occurs at 180°C. Then, the formation of Na_3AlH_6 is observed and its further decomposition into NaH and Al.

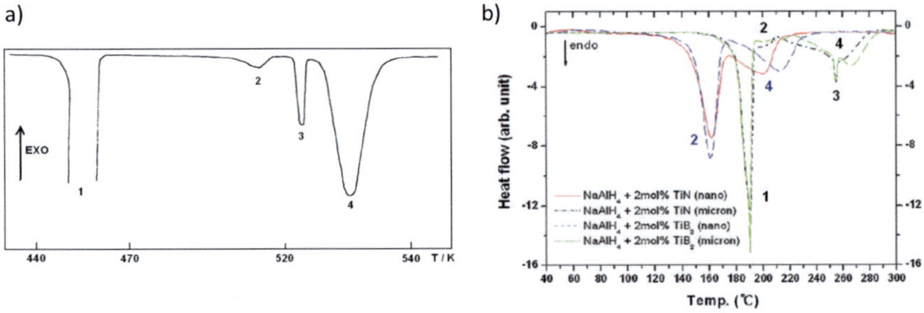

Figure 1.6: DSC traces for a) undoped $NaAlH_4$ (from [49]) and b) for doped $NaAlH_4$ (from [60]). 1: melting of $NaAlH_4$, 2: $NaAlH_4 \rightarrow Na_3AlH_6$, 3: α-$NaAlH_4 \rightarrow$ β-Na_3AlH_6 and 4: $Na_3AlH_6 \rightarrow NaH$

For doped-$NaAlH_4$, the sequence observed by DSC is slightly different (**Figure 1.6b**) with only two endothermic peaks arising below 300°C. The decomposition of $NaAlH_4$ [reaction (1.6)] occurs below the melting point (150-170°C). The second step reflects the decomposition of Na_3AlH_6 [reaction (1.7)] before its structural transition (180-220°C) [61]. The addition of $TiCl_3$ clearly improves the kinetic properties of $NaAlH_4$ [62, 63]. Kircher et al. [64] have determined an activation energy for $NaAlH_4$ doped with nanocrystalline Ti and other Ti-based compounds of *ca.* 100 kJ mol^{-1}. For the decomposition of Na_3AlH_6, they have found a higher activation energy (190-200 kJ mol^{-1}), which is in agreement with the lower reaction rate observed for the second decomposition step [eq. (1.7)] [65]. Sandrock et al. [63] have shown that higher reaction rates are obtained with increasing the dopant content but with the decrease of the hydrogen capacity as a counterpart. By using the Arrhenius law for modelling the reaction kinetics, they have found that only a small amount of dopant is necessary to decrease the thermal activation. A good compromise between high rates and capacities is 4mol% $TiCl_3$ added to sodium alanate.

1.2.1.4. Synthesis

The synthesis of sodium alanate from NaH and using different solvents is known for quite a long time [66]. The original study by Bogdanovic and Schwickardi [18] has involved a complicated chemical multi-step process, which has been enhanced afterwards by using ball milling. Better H_2 sorption performances have been described for $NaAlH_4$ prepared by solvent mediated milling from NaH and Al in the presence of a Ti dopant [67]. Afterwards, even better results have been obtained with dry milling of the same elements; the milling process especially improves the doping efficiency [51, 68-73]. During ball milling, the diminution of grain and particle sizes leads to shorter reactions paths and promotes fast diffusion. The direct preparation of sodium alanate from NaH and Al by reactive milling under H_2 atmosphere [74-76] is a simple method for the synthesis of doped-$NaAlH_4$. Bellosta von Colbe et al. [76] have managed to synthesise doped-

$NaAlH_4$ in one step by milling NaH, Al and $TiCl_3$ under ~80 bar H_2. They could follow the reaction by recording temperature and pressure in the vial during milling. The hydrogen pressure decreases during milling because of the absorption of hydrogen and then the reaction is completed after 4 h. The XRD patterns after milling show the presence of $NaAlH_4$, Al and NaCl and the capacity of the first desorption is quite high (almost 4 wt% H_2). Doped-$NaAlH_4$ prepared by the one-step synthesis method shows enhanced kinetic properties compared to $NaAlH_4$ milled with the same dopant but in a separate doping step [77].

Scanning electron microscopy (SEM) and transmission electron microscopy (TEM) studies [78, 79] have shown a homogeneous distribution of Ti species within the powder after ball milling. After the first desorption, the particles become very porous. In addition, the morphology of the particles is significantly changed after desorption. Nevertheless, no cracks or pulverization of the particles are observed after desorption. This is a major difference from the results obtained for intermetallic hydrides (e.g. MgH_2). However, an even short exposition to air of the sample during transfer to the SEM chamber strongly changes the morphology of the powder. In addition, the material is very sensitive to the electron beam that promotes the decomposition of the hydride phase. The homogeneous dispersion of the dopant is an important feature, explaining the strong influence of the doping method [62]. The re-hydrogenation reaction may start at the interface between Al and NaH initially, where first Na_3AlH_6 is formed and afterwards reacts to obtain $NaAlH_4$. It is thus necessary to have a good distribution of all elements otherwise the core of the Al particle remains unreacted and the reaction is not complete.

1.2.1.5. Addition of dopants

As already discussed, the use of a dopant greatly improves the kinetics of H_2 absorption and desorption. Initially, Ti or Zr alkoxides have been used as dopants but chloride compounds are now preferred, because the alkoxides are heavy compounds, which decrease the total capacity of the hydride and lead to significant hydrocarbon and oxygen contaminations [80]. In the last years, a wide range of compounds have been tested to improve the performance obtained with $TiCl_3$ proposed by Bogdanovic and Schwickardy in 1997 [18]. For example, Lee et al. [61] have studied the effect of lanthanide oxides as dopants. The main motivation for the choice of oxides is that they do not react with the alanate to form inert compounds (e.g. NaCl). All the oxides have not the same efficiency but they all lead to a decrease of the reaction temperature and an increase of the kinetics. However, the capacity fading during several absorption-desorption cycles is the same as for other dopants.

Anton [65] has analysed the dehydrogenation rates for doped-$NaAlH_4$ prepared by milling $NaAlH_4$ together with different chloride compounds MCl_x (M: mainly transition metals and x=2, 3 or 4) for 15 min (**Figure 1.7**). For several compounds, the desorption rates are in the same range as for the as-milled undoped sample (Type III). The highest rates are obtained for Ti^{3+} and Ti^{4+} cations (Type I) but with a maximum hydrogen capacity of 3 wt% H_2 for 2 mol% dopant. In

Anton's study, the anion species appear to have no influence on the desorption properties, e.g. the same decomposition rates are obtained for $NaAlH_4$ doped with $TiCl_3$ or TiF_3 (**Figure 1.7a**). However, in recent papers [81-83], it is shown on the contrary that TiF_3 could be a better dopant than $TiCl_3$.

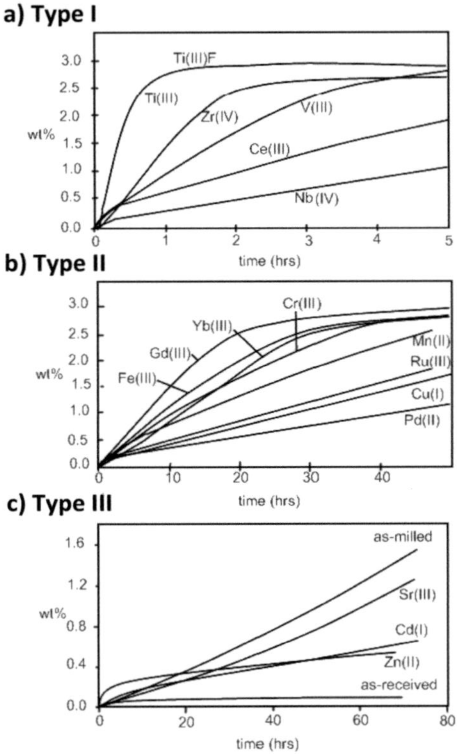

Figure 1.7: Dehydrogenation curves of doped-$NaAlH_4$ with *a)* highly active chloride dopants (type I), *b)* moderately active chloride dopants (type II) and *c)* inactive chloride additives (Type III). Figure taken from [65] (not all chlorides are shown here)

Alternatively, different groups [84, 85] have tried to dope $NaAlH_4$ with metallic Ti. The main idea is to avoid the formation of by-products like NaCl and then obtain a higher gravimetric capacity. Enhanced reaction rates are obtained for sodium alanate milled with Ti under H_2 atmosphere compared to $NaAlH_4$ milled alone under the same conditions. However, the reaction kinetics are not as fast as with titanium chloride, especially for the second step of decomposition [eq. (1.7)].

Moreover, the hydrogen capacity decreases with cycle number.

From most of the studies published up to now, it appears that the most efficient dopants are $TiCl_3$ and nanoclusters of titanium Ti_{13}. A problem with these last ones is the difficulty to synthesise them [86, 87]. The hydrogen sorption rates are higher than for $TiCl_3$, likely due to a better dispersion of the Ti species and to size effects. The use of such small particles can decrease the absorption time to very short ones (10-15 min). Very small dopants enable the achievement of a higher dispersion of the catalytically active Ti species, which probably facilitates the mass transfer. Unfortunately, this effect disappears during cycling.

Very recently, a new set of dopants ($ScCl_3$ and $CeCl_3$ mainly) has been reported to have a higher efficiency than $TiCl_3$ [77, 88-91]. $ScCl_3$ has a similar effect as $TiCl_3$ on the reaction temperature for both decomposition steps, but with a lower capacity loss during cycling. In addition, with Sc-based dopant, the kinetic properties after cycling are better than using Ti-based dopant. Wang et al. [88] have linked this improvement upon cycling to the decrease of the activation energy after cycling whereas it increases with $TiCl_3$. The kinetics in both cases highly depend on the amount of dopant. Doping with Sc compounds may lead to a more delocalised structure (i.e. H atoms "move" from the hydride to Sc), which reduces the desorption energy [90]. It is also suggested that the formation of $ScAl_3$ or $ScAl_5$ compounds after dehydrogenation has a positive effect on the kinetics, similarly to Ti-Al alloys (see below) [90]. However, the desorption energy for the release of H_2 is slightly higher than for Ti-Al compounds and this suggests that for desorption. Ti-based dopants are better than Sc-based dopants [92, 93]. The presence of $ScAl_3$ has been observed by NMR spectroscopy [94] and results from the decomposition of $ScCl_3$ during milling. Nevertheless, other phases like remaining $ScCl_3$, $Al_{1-x}Sc_x$ and ScH_x have also been detected. In addition, a memory effect (decrease of the capacity if not fully de-/absorb) is found by using $ScCl_3$ as dopant and this limits its cycling properties when using different sorption conditions. This memory effect has not been observed using $CeCl_3$ or $PrCl_3$. In addition, fast hydrogenation has been obtained using $CeCl_3$. This sample shows very good cycling properties with still 4 wt% H_2 capacity after 95 sorption cycles [89].

1.2.1.6. Catalytic mechanism

Considering that $TiCl_3$ has been the first very efficient dopant to be discovered, many papers speculate on the exact role of Ti in doped sodium alanate and on the catalytic mechanism. The results obtained are described here. Several different analysis techniques have been used to investigate the role of titanium [78, 86, 95-100]. The only consensual result is the change of chemical state (from Ti^{3+} to Ti^0) of titanium reducing to the metallic state during milling [78, 100]. This is related to the following reaction:

$$3NaAlH_4 + TiCl_3 \rightarrow 3NaCl + 3Al^0 + Ti^0 + 6H_2 \quad \textbf{(1.9)}$$

This reaction consumes Na and Al and then limits the reverse reaction of absorption and decreases the capacity during cycling. Léon et al. [95] have performed XPS analyses on sodium alanate doped with $TiCl_3$. They have confirmed that titanium is in the metallic state after doping and they have shown that the amount of titanium decreases at the surface by increasing the milling time. The surface may be covered by NaCl [from eq. (1.9)] and Al_2O_3. The contamination by oxygen is only a surface effect and no more Al_2O_3 is detected below the surface. Most of the aluminium is also in the metallic state. With X-ray absorption analyses, different groups [98-100] have given other evidences of the metallic state of Ti after milling. It has also been shown that Ti substitutes for Al in the alanate and then increases the reaction rate [97]. As $TiCl_3$ is a good dopant, better than pure Ti for example, one can assume that the reduction to the zero-valent state plays an important role for the catalytic activity of Ti species. Gomes et al. [101] have performed vibrational spectroscopy studies (Raman and IR) to investigate the titanium effect. They have detected the presence of titanium in the local environment of $[AlH_4]^-$ units supporting the idea of titanium diffusion and substitution in the alanate lattice, which leads to enhanced kinetic properties. However, another Raman spectroscopy study [102] indicates that the $[AlH_4]^-$ anion is very stable even under melting. Then, the role of titanium could be to help to break this structure (e.g. by weakening the Al-H bond).

Other authors [100, 103-111] have proposed that Al-Ti compound(s) (mainly Al_3Ti), formed during milling or cycling, is the active species rather than Ti. This idea is supported by Brinks et al. [105], who have reported the formation of a Ti-based amorphous compound during doping and then the formation of a Ti-Al phase after several hydriding/dehydriding cycles. DFT calculations show that the formation of Al_3Ti is highly favourable from the thermodynamic point of view. The easiest route for the formation of Al_3Ti involves the formation of AlH_3 by decomposition of $NaAlH_4$ and then the reaction of AlH_3 with metallic Ti to form Al_3Ti [110]. Haiduc et al. [106] have observed by XRD a small broad reflection close to the Al main peak after adding $TiCl_3$ as dopant. The authors have suggested the formation of an hcp-Ti(Al) solid solution, which may be favoured instead of TiH_2 when Ti and Al are present and when metallic Al is produced according to equation (1.9). By increasing the temperature, the formation of a solid solution leads to a more ordered Al_xTi compound. Two other groups [98, 100] have shown the formation of Al_3Ti by X-ray absorption analyses. Dobbins et al. [112] have found that $TiCl_3$ decomposes during the very beginning of the milling to form metallic Ti. However, the formation of Al_3Ti occurs only after longer milling time by the reaction of Ti metal with Al in $NaAlH_4$. On the contrary, by TEM analysis, Graham et al [113] have shown a progressive change of the Ti chemical state from $TiCl_3$ to Al_3Ti during cycling. To check if Ti-Al compounds are indeed the active species, Resan et al. [73] have directly added Al_3Ti or Ti_3Al as dopant to $NaAlH_4$. They have observed a small decrease of the total hydrogen capacity and a decrease of the reaction temperature of the same order as for a sample milled without dopant. This tends to confirm that the formation of an Al-Ti compound is not the critical point of the catalytic mechanism and that the decomposition of $TiCl_3$ itself in the early milling process plays also an important role. This difference may be explained by the beneficial influence of the formation of Ti-Al-H complexes during milling rather than intermetallic compounds [92].

According to Bellosta von Colbe et al. [114], a clear effect of the titanium species is to dissociate the H_2 molecule and facilitate the transport of H atoms into the bulk. This has been confirmed by a theoretical study of Ti at an Al surface [115]. However, there is most probably another effect responsible especially for improving re-hydrogenation. Bellosta von Colbe and coworkers have shown that this effect should be related to a mass transfer process, necessary for reacting NaH and Al, which are segregated after the decomposition step.

Different groups [59, 109, 116] have reported that the main mechanism involved in the hydriding/dehydriding reaction is the diffusion of alane AlH_3. The presence of this compound allows to explain all the steps of sorption in sodium alanate in a simple manner. For example, 3 NaH and AlH_3 react to form Na_3AlH_6. If AlH_3 is the main active element, the role of Ti in doping could be explained by the formation of Ti-Al compounds that facilitates the alane formation and H_2 sorption. It has also been reported that Ti favours the dissociation of alane [117]. The presence of alanes has been reported experimentally by Fu et al. [118], who have detected the presence of AlH_3 by Inelastic Neutron Scattering (INS).

Additionally, recent publications consider the formation of defects in sodium alanate that influence the diffusion process inside the material and then the kinetic properties [119-123]. Shi et al. [124] have found that hydrogen diffusion is H-vacancies mediated. For Na_3AlH_6, Li et al. [125] have reported that the creation of Na vacancies is energetically more favourable than the creation of Al vacancies because it is not necessary to break any Al-H bonds. Moreover, the Na vacancies can efficiently promote hydrogen release because they are more efficient to decrease the necessary energy. The formation of NaCl could be the signature of vacancies, as shown for example by Vegge [123], who has found in addition that the formation of Na vacancies is favoured by the presence of metallic Ti. From DFT calculations, Lodziana et al. [126] have found that Ti is positively charged and thus attracts the negatively charged defects like Na vacancies. Palumbo et al. [127-130] have shown that the defects formed in $NaAlH_4$ are H-vacancies in Na_3AlH_6 and the number of these defects increases when the Ti catalyst is added. The H-vacancies and the Ti sites act as traps for H that diffuses by hopping between these traps. Similarly, Singh et al. [131] have reported an increase of the number of H-vacancies when adding $TiCl_3$ to $NaAlH_4$. Besides, they have demonstrated that the crystallite size in $NaAlH_4$ stays small when $TiCl_3$ is added. The combination of these two effects may greatly improve the diffusion mechanisms inside the compound and then improve the kinetics. The reduced crystallite size could be related to the presence of grain refiners appearing during decomposition: Al_xTi compound for the Al phase and NaCl for NaH. Gunaydin et al. [132] have demonstrated by first-principles DFT calculations that the limiting step for the decomposition of $NaAlH_4$ is the diffusion of Al through the AlH_3 vacancies because the energy involved is similar to the experimental activation energy. Once AlH_3 has reached the surface of $NaAlH_4$, Ti species may catalyse the breaking of H-Al bonds and facilitate the nucleation of the Al phase by the formation of Al_3Ti (**Figure 1.8**).

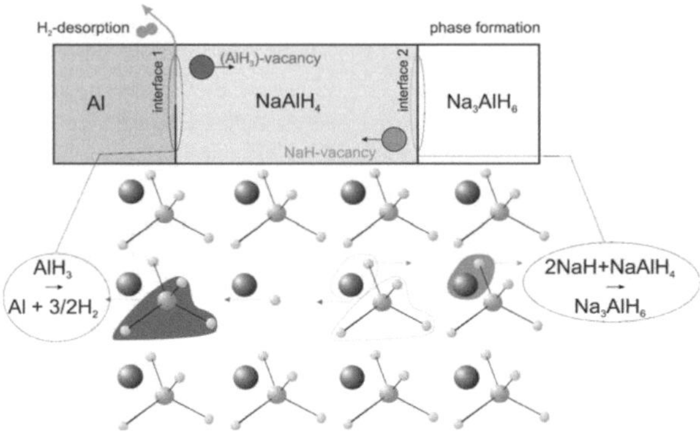

Figure 1.8: Sketch of $NaAlH_4$ decomposition via vacancy diffusion [120] (adapted from [132]).

Despite the very different results reported on the role of titanium, there is now a growing consensus about the role of vacancies or defects. They may be created during the decomposition of $TiCl_3$ and the formation of NaCl that explains why metallic Ti or Al-Ti alloys are not as efficient dopants. The vacancies can favour the diffusion of hydrogen or hydrogenated species. The diffusion of AlH_3 is a possible reaction route but there is a lack of experimental proof. At present, the analysis of the role of the dopant changes to the investigation of the other compounds, such as $ScCl_3$ or $CeCl_3$, found to have an equal or even better efficiency than $TiCl_3$. Few works study the role of dopants like $ScCl_3$ or $CeCl_3$, although they lead to enhanced kinetic properties. Bogdanovic et al. [89] have shown very fast H_2 absorption kinetics, which they have attributed to the formation of Sc- or Ce- hydrides. For $ScCl_3$ used as dopant, only theoretical results have been published [92, 93]. These calculations have shown that the formation of TM-Al-H complexes (TM for transition metal) help the release of H_2 by reducing the hydrogen desorption energy [93] but the energy is lower for Ti-Al-H compounds than for Sc-Al-H compounds. Nevertheless, a particular interest in Ti or Sc is that they also promote the absorption reaction in a similar way through TM-Al-H complexes by the facilitated exchange of electrons [92].

Wan et al. [133] have studied the chemical state of Ce in a $NaAlH_4$ sample doped with CeH_2. They have found the formation of Al-Ce compounds after milling and also that some Ce ions are substituted for Na in the alanate structure. The chemical state of Ce has been analysed by XANES experiments and the authors have found that Ce has the oxidation degree of +IV after milling, which changes to +III after cycling the sample. Using XAS (X-ray absorption spectroscopy), Léon et al. [134] have shown that $CeCl_3$ remains as a chloride compound after milling and cycling with $NaAlH_4$. They propose that the better cycling properties obtained by using this dopant may be related to the absence of the decomposition of its dopant during milling. On the contrary, Fan et al. [135] have shown that the formation of Al_4Ce during cycling from the decomposition of $CeCl_3$ is the real catalytic species for the release/uptake of hydrogen in $NaAlH_4$. They have found similar reaction rates for the hydriding reaction of a mixture of NaH and Al doped with $CeCl_3$ or Al_4Ce. In addition, when using Al_4Ce, there is no by-products, like NaCl, formed in the sample and the capacities achieved are higher than with $CeCl_3$ as dopant. The catalytic mechanism is not described in detail, in particular not for the dehydriding reaction. The first results reported for $ScCl_3$ and $CeCl_3$ as dopant are quite different to what has been reported for the Ti-based dopant and highlight the necessity to study in more detail these other dopants to understand the reaction and catalytic mechanisms in $NaAlH_4$. This is the subject of the first part of this work.

1.2.2 $Ca(BH_4)_2$: a high capacity storage material

Metal tetrahydroboranate or metal borohydride compounds with the general formula $M(BH_4)_n$ (M=Li, Na, Ca, Mg,...) are more attractive than alanates for storage because of their very large capacity, e.g. 18.5 wt% H_2 for $LiBH_4$ [19, 136]. However, they suffer from very strong limitations: the kinetics are very poor and reversibility is difficult to achieve. Nevertheless, these compounds are considered as very promising and the activities devoted to these compounds are increasing quickly.

$LiBH_4$ is considered as a reference for metal borohydride. It displays the highest possible hydrogen capacity of all borohydrides. The decomposition is often described as a one-step reaction according to equation (1.10), involving the release of 13.5 wt% hydrogen [37, 137]. The full capacity is reached after the decomposition of LiH occurring at very high temperature and generally is thus not considered for practical applications.

$$LiBH_4 \rightarrow LiH + B + 3/2\,H_2 \quad (13.5\text{ wt\% } H_2) \qquad \textbf{(1.10)}$$

This compound has been intensively studied in the last years and some authors have proposed a decomposition of $LiBH_4$ in several steps with the formation of intermediate compounds such as $Li_2B_{12}H_{12}$ [26, 136, 138-140]. This has been confirmed by a recent NMR study of $LiBH_4$ during decomposition [141]. For pure $LiBH_4$, reaction (1.10) proceeds at a temperature higher than 400°C. Its reaction enthalpy is high and prevents a practical use for mobile application. The enthalpy has been experimentally determined to 74 kJ mol^{-1} H_2 [29]. The reverse reaction needs very drastic conditions (p>100 bar and T>350°C) [142]. In addition, one drawback of $LiBH_4$ is the release of very toxic diborane B_2H_6 during decomposition as a concurrent reaction to the release of hydrogen [136]. Therefore, other metal borohydrides, $Mg(BH_4)_2$ and $Ca(BH_4)_2$, have been considered recently. $Ca(BH_4)_2$ is chosen here because of its particularly interesting thermodynamic properties.

1.2.2.1. Hydrogen sorption properties of $Ca(BH_4)_2$

$Ca(BH_4)_2$ has a high hydrogen content (11.6 wt% H_2) and a low predicted enthalpy for reaction (1.11) proposed for its formation/decomposition (32 kJ mol^{-1} H_2 [31] and 41.4 kJ mol^{-1} H_2 [42]). The total hydrogen release for this reaction is theoretically 9.6 wt% H_2.

$$3Ca(BH_4)_2 \leftrightarrow 2CaH_2 + CaB_6 + 10H_2 \quad \text{(9.6 wt\% H}_2\text{)} \qquad \textbf{(1.11)}$$

The predicted range of reaction enthalpy for $Ca(BH_4)_2$ (30-40 kJ mol^{-1} H_2) is in the target proposed to suit the DOE requirements for mobile application (**Figure 1.9**). This implies a decomposition temperature in the 100-150°C range at 1 bar H_2 as for $NaAlH_4$. A similar enthalpy value (40.8 kJ mol^{-1} H_2) has been reported by Ozolins et al. [41] but they also found a similar value (39.2 kJ mol^{-1} H_2) for the decomposition of $Ca(BH_4)_2$ in $CaB_{12}H_{12}$ and CaH_2 (reaction (1.12)). This result indicates that both reactions are competing. Nevertheless, $CaB_{12}H_{12}$ has seldom been reported as decomposition product of $Ca(BH_4)_2$ (see below §1.2.2.4).

$$6Ca(BH_4)_4 \rightarrow CaB_{12}H_{12} + 5CaH_2 + 13H_2 \quad \textbf{(6.3 wt\% H}_2\textbf{)} \qquad \textbf{(1.12)}$$

The lower decomposition temperature of $Ca(BH_4)_2$ compared to $LiBH_4$ has also been predicted by Nakamori et al. [35] from the electro-negativity of calcium.

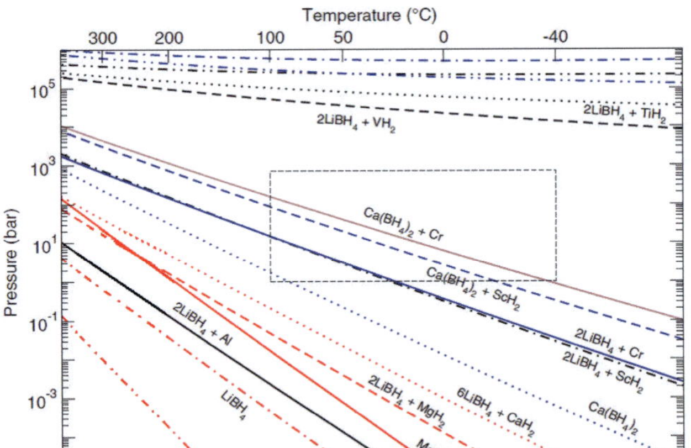

Figure 1.9: van't Hoff plots for different reactions calculated by Siegel et al. (reproduced from [42]). The dash-line box represents a range of pressure and temperature desirable for hydrogen storage. The dotted blue line for $Ca(BH_4)_2$, decomposing in CaB_6 + CaH_2 crosses this range.

However, the decomposition of $Ca(BH_4)_2$ has experimentally been found to occur in two steps (§ 1.2.2.4). The first report on the experimental determination of the thermodynamic properties of $Ca(BH_4)_2$ has been published in 2010 [143]. The enthalpy value has been determined by measuring the PCI curves at different temperatures for the first decomposition step (6-7 wt% H_2 released). The enthalpy derived from the van't Hoff plot is 87 kJ mol^{-1} H_2 and is much higher than calculated previously. However, the authors have also reported sluggish reaction kinetics and this value is probably overestimated as usually when taking into account only the desorption step (see §2.3). The determination of the absorption step is necessary to give the lower and upper values where the true enthalpy value lies.

The decomposition of pure $Ca(BH_4)_2$ appears to follow a different route than for $LiBH_4$ with the formation of CaB_6 (and not pure B) as decomposition product. The formation of CaB_6 could be an advantage since the formation of metal borohydride from the metal hydride and elemental boron has been shown to be very difficult [142].

1.2.2.2. Crystal structure

$Ca(BH_4)_2$ presents different polymorphs, which are observed depending on the synthesis method or the applied heat treatment (see **Annexe B**). These different structures make the analysis of $Ca(BH_4)_2$ by XRD very complicated, especially because several structures are stable at room temperature. The first calculated structure of $Ca(BH_4)_2$ at room temperature is the orthorhombic (*Fddd*) α-phase [31, 144, 145]. However, this phase has not been observed by Riktor et al. [146] after chemical synthesis of $Ca(BH_4)_2$, but they have reported the presence at room temperature of two other phases: γ-$Ca(BH_4)_2$ (orthorhombic) and β-$Ca(BH_4)_2$ (tetragonal). They have also reported the presence of a fourth phase called δ-$Ca(BH_4)_2$ (**Figure 1.10**). Later studies [147, 148] have given the crystal structure of β-$Ca(BH_4)_2$ with space group $P4_2/m$ and the structure of the γ-phase with space group *Pbca* [145, 148-150]. These different phases can be distinguished by Raman spectroscopy [147].

Different space groups have been proposed for the different polymorphs [149, 151]. In addition, the structure of another polymorph α' with space group *I-42d* has been reported and is observed after a second order transition of the α phase at 222°C. The α and α' phases transform into the β phase above 180°C. The β phase is the only observed phase for temperatures higher than 297°C; it then decomposes between 382 to 387°C into unknown compound(s). In addition, the α phase is very stable and is only deformed when high pressures are applied, whereas the β phase disappears above 10 GPa to form a highly disordered phase that has not been identified yet [152].

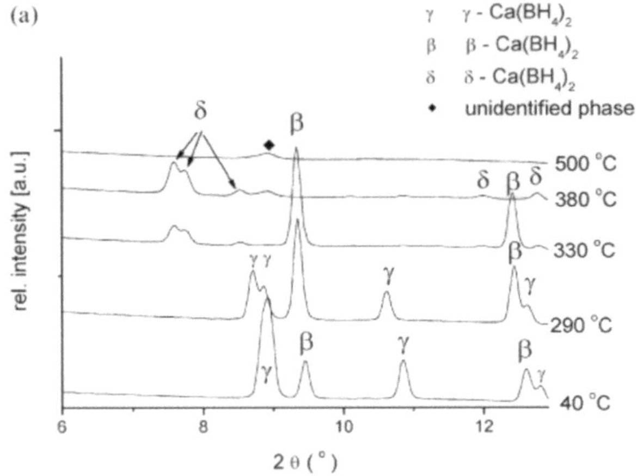

Figure 1.10: Different $Ca(BH_4)_2$ polymorphs as observed by in-situ synchrotron measurements by Riktor et al. [146] during heating of this material.

1.2.2.3. Synthesis

The synthesis of $Ca(BH_4)_2$ is difficult despite its high thermodynamic stability. One of the first used methods was the metathesis reaction [153] from $LiBH_4$ (or $NaBH_4$) and $CaCl_2$. But LiCl is formed as by-product and should be separated from the borohydride in a second step. Another synthesis method is by wet chemical synthesis. This method necessitates the use of a large quantity of solvent, which has to be removed in additional purification steps [154]. Rönnebro and Majzoub [155] have demonstrated the possibility to synthesise $Ca(BH_4)_2$ without by-products from the reaction between CaH_2 and CaB_6 under 700 bar H_2 and 440°C for 48 h. The reaction is possible by adding a mixture of $TiCl_3$ and Pd additives. With 8 wt% of these additives, the reaction yields 60% of $Ca(BH_4)_2$. Similar yields have been reported by Kim et al. [44] for a mixture of CaH_2 + CaB_6 + Mg produced by the decomposition at 400°C of $Ca(BH_4)_2$ +MgH_2 that re-forms these two hydrides after 48 h at 350°C and 90 bar H_2, but with the remaining intermediate phase $Ca_4Mg_3H_{14}$. Nevertheless, the authors have shown that the yield strongly depends on the decomposition and hydrogenation conditions (e.g. intermediate milling step to create fresh surfaces, temperature, time).

1.2.2.4. Reaction paths and effect of dopants.

Few studies are published on the hydrogen sorption properties of $Ca(BH_4)_2$ because of its novelty for hydrogen storage application. The DSC trace of $Ca(BH_4)_2$ under Ar shows three endothermic peaks [155, 156] (**Figure 1.11**). The first at 140-170°C is related to a structural transformation of $Ca(BH_4)_2$ from the α to the β-phase. The two following peaks (starting at ca. 350-360°C and 450-

500°C) are related to the desorption of hydrogen in two steps with a total weight loss close to 9 wt% H_2. The two decomposition steps involve first the formation of CaH_2 and an intermediate phase and then the decomposition of the intermediate compound into CaH_2 and CaB_6 (or amorphous B) [156, 157].

Aoki et al. [157] have indexed the intermediate phase as an orthorhombic structure. However, the true structure of this phase is still not known. The decomposition of $Ca(BH_4)_2$ in this phase releases 5.9 wt% of hydrogen and the PCI analysis shows a plateau at 6 bar at 320°C. It is possible that only a part of the $Ca(BH_4)_2$ decomposes into this intermediate since CaB_6 is detected by Raman spectroscopy already after the first decomposition step [158]. The formation of an intermediate phase $CaB_{12}H_{12}$ has also been proposed by analogy with recent works proving the presence of $M_{2/n}B_{12}H_{12}$ compounds during the decomposition of several borohydrides [140, 141, 159]. However, no unambiguous trace of this compound has been reported experimentally up to now. Riktor et al. [160] have studied the decomposition of $Ca(BH_4)_2$ by ex-situ XRD by stopping the heating ramp at different temperatures. The results are different from those for in-situ experiments. They have observed the formation of a new phase between 370 and 400°C after a first release of hydrogen. This intermediate phase is identified as CaB_2H_x (likely x=2, see **Annexe B**) and decomposes above 400°C into CaH_2 and CaB_6. They have also found the formation of amorphous boron or a boron containing phase which can limit the reversibility of the reaction.

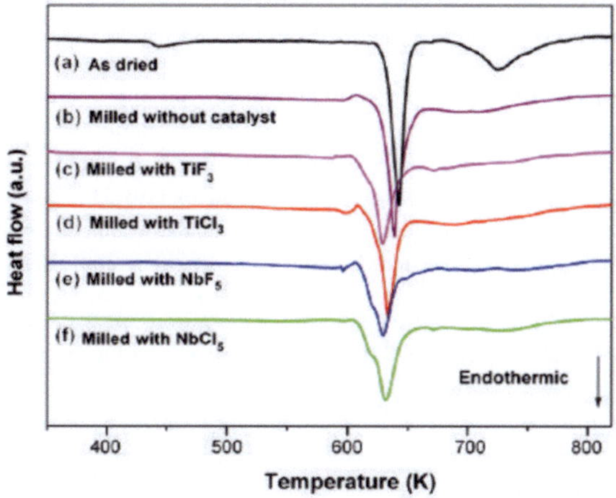

Figure 1.11: DSC traces (obtained by Kim et al. [161]) during heating of $Ca(BH_4)_2$ milled with different additives.

The use of additives like $TiCl_3$ allows a small decrease of the desorption temperature (**Figure 1.11**) [162], but the formation of by-products (e.g. $CaCl_2$) during milling with $Ca(BH_4)_2$ decreases the total hydrogen capacity. However, 50-60% of re-hydrogenation is obtained under 90 bar H_2 and 350°C for 24h. Kim et al. [161] have tried different additives to improve the kinetic properties of pure $Ca(BH_4)_2$. The effect of TiF_3, $TiCl_3$, NbF_5 or $NbCl_5$ is small for the decomposition reaction because only a small decrease of the decomposition temperature is observed that is mainly obtained by the milling process itself (**Figure 1.11**). By contrast, the re-absorption is only possible by using an additive. The best result is obtained using NbF_5 : 5 wt% H_2 is absorbed with the highest rate under 90 bar H_2 and 350°C for 24 h. The authors have indicated that the better kinetics obtained with NbF_5 may be related to the low melting point (77°C) of this compound that allows a better dispersion in $Ca(BH_4)_2$ during milling because NbF_5 is in the liquid state.

Wang et al. [159] have used different additives (Nb_2O_5, $RuCl_3$ and $TiCl_3$) with little effect on the kinetics and the reversibility of reaction (1.11). They have also found the formation of $CaB_{12}H_{12}$. From their results they have claimed that $Ca(BH_4)_2$ is not a suitable compound because of the formation of the very stable $CaB_{12}H_{12}$ and the poor kinetics. However, their conclusion is questionable. First, their doping method appears to be unsuccessful: they could not achieve a good dispersion of the dopant by ball milling whereas this technique is widely and commonly used to obtain a good distribution of additives in different materials. In addition, it has been shown previously by Kim et al. [161] that the fluoride compounds are better dopants than chloride compounds so there is still space for improvement in the kinetic properties of $Ca(BH_4)_2$, in particular for hydrogenation. Kim and coworkers have also clearly shown that $Ca(BH_4)_2$ is obtained after rehydrogenation and no trace of $CaB_{12}H_{12}$ has been found. The conclusion from Wang and co-authors is then probably misleading due to unsuitable experimental conditions. Nevertheless, at the time of writing, the formation of $CaB_{12}H_{12}$ during decomposition cannot be ruled out completely and may be a competing reaction to the formation of CaB_6 and CaH_2. The experimental results reported so far suggest a low yield of $CaB_{12}H_{12}$, difficult to identify by conventional characterisation techniques. The presence of this compound may depend on the experimental conditions (atmosphere, temperature) as demonstrated for $LiBH_4$ [45].

In a slightly different approach, Lee et al. [163] show a potential interest of using composites of borohydride, here x $LiBH_4$ + $(1-x)$ $Ca(BH_4)_2$. They have reported the reaction between the two compounds during decomposition to form LiH and CaB_6. In addition by DSC, the authors have observed a eutectic melting at ca. 200°C rather than the melting of pure $LiBH_4$ at 280°C. The lowest decomposition temperature (without additive) is obtained for x=0.4. The partial reversibility is obtained by using NbF_5 as additive.

1.2.3 Open questions for complex hydrides and aim of this work

Sodium alanate is, at the time of writing, the only complex hydride showing reasonable operating temperature and H_2 release/uptake rates. It has been widely studied and the H_2 sorption properties have been greatly improved by the addition of a dopant. The most commonly used dopant is $TiCl_3$, but more recently proposed $ScCl_3$ and $CeCl_3$ appear to improve further the reaction rates, in particular for the absorption step. However, despite intensive studies of the catalytic mechanism, in particular for $TiCl_3$-doped $NaAlH_4$, there is still no clear understanding of the process. In particular, the previous studies have been mainly focused on the Ti role, whereas several mechanisms are likely involved. In addition, only few works have been reported on the role of the other possible dopants $ScCl_3$ and $CeCl_3$. The first reports suggest a different behaviour of these compounds compared to $TiCl_3$ and this result needs to be clarified. The aim of this work is then to study differently-doped $NaAlH_4$ samples that will be prepared by reactive ball milling, a particularly suitable technique for this type of compounds. The role of the different dopants on the thermodynamic and kinetic properties of $NaAlH_4$ will be analysed in detail. From the results already published, it is likely that different behaviours will be observed using different dopants. Comparing the samples doped with Ti, $TiCl_3$, $ScCl_3$ and $CeCl_3$ should give new information about the catalytic mechanism. The aim of this work is thus to achieve a global understanding of this mechanism that is a very important issue to further improve the kinetics of $NaAlH_4$ for tank applications.

The main drawback of $NaAlH_4$ for hydrogen storage is the relatively small hydrogen capacity (4 wt% H_2) experimentally achieved, which is well below the requirements usually considered for mobile applications. Therefore, metal borohydrides have attracted lot of interest in the last years because of their very high hydrogen capacities (>10 wt% H_2). However, these compounds are very stable and then not useful for tank applications; they are also difficult to investigate experimentally. From calculations, $Ca(BH_4)_2$ has been found to have suitable thermodynamic properties, similar to $NaAlH_4$. Initial works published have shown the possible reversible H_2 release/uptake from/in this compound but with sluggish kinetics despite the addition of different additives. The decomposition appears to proceed in several steps but the intermediate compounds formed at each step are not well defined. The very slow reaction rates hinder an experimental determination of the thermodynamic properties, in particular the reaction enthalpy for hydrogen absorption/desorption. In fact, no complete experimental confirmation of the calculated enthalpy for desorption and absorption has been reported so far. The high decomposition temperatures reported in literature are then either related to a higher reaction enthalpy than expected or to slow kinetics. The first experimental determination of the enthalpy gives a high value but has to be confirmed because only the desorption step has been considered. In addition, no simple and efficient synthesis method has been proposed, whereas it is an important point for an up-scaling of the production of $Ca(BH_4)_2$. Reactive ball milling is an alternative to wet chemistry methods and its feasibility will be studied here. The influence of different additives will be analysed in detail for the synthesis but also for further cycling in order to improve the reactions rates. As for $NaAlH_4$, the comparison of different additives helps the understanding of the decomposition and re-hydrogenation mechanisms as well as the role of the additive to improve the H_2 sorption properties. The improvement of the kinetics is of crucial importance for the development of solid state hydrogen storage for mobile applications.

2. Experimental

Materials and methods

Reactive ball milling is used in this work as a synthesis method for all samples. This section describes the preparation of hydride samples by this technique as well as the other characterisation methods used for the structural and thermal analyses of the materials. The structural analysis techniques were use to determine the structure of the complex hydrides, intermediate phases and also of the additives used. The thermal analysis is focused on the determination of properties relevant for hydrogen storage, such as decomposition temperatures and enthalpies of reaction. All powder handling is done under controlled atmosphere, mainly in a glove box with Ar atmosphere at oxygen and water levels below 1 ppm.

2.1 High pressure reactive ball milling

Reactive ball milling was used for the synthesis of $NaAlH_4$ and $Ca(BH_4)_2$. This method is usually very simple and the synthesis can be done in one step. The conventional wet chemical methods involve several steps and require a large quantity of solvent that has to be removed afterwards. Reactive milling is also a very interesting method because it allows the preparation of pure or doped hydrides. The milling is performed under a reactive atmosphere, here hydrogen, that participates directly in the reaction. For the synthesis of hydrides, the use of high hydrogen pressures has been shown to favour the hydriding reaction [164]. A specially designed milling vial was used that is operating with up to 150 bar H_2 (evico magnetics) (**Figure 2.1**). In addition, this vial is equipped with pressure (p) and temperature (T) sensors that are placed in the lid and allow to monitor p and T during milling. The electronic system placed in the lid sends the data to a computer through an antenna and a receiver. This gives direct information about the reactions occurring during milling. The autonomy of the batteries used for the sensors allows milling for maximum 25 h without interruption. The vial is installed in a planetary monomill Fritsch Pulverisette P6.

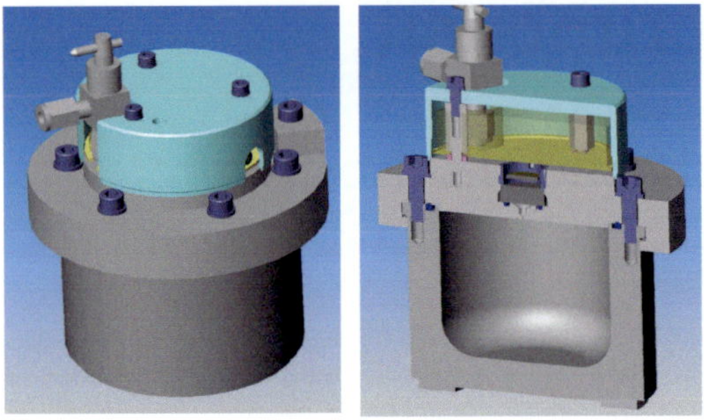

Figure 2.1: Scheme of the milling vial used for reactive ball milling (evico magnetics).

For the synthesis of $NaAlH_4$, the starting materials were NaH (Sigma Aldrich, 95 %) and Al (MaTeck, -100+200 mesh) (in ratio 1:1) with dopants of different nature and in different quantities (**Table 2.1**). The dopants were $TiCl_3$ (Sigma Aldrich, 99.999 %), $ScCl_3$ (Sigma Aldrich, 99.9 %), $CeCl_3$ (Alfa Aesar, 99.9 %) or Ti (Alfa Aesar, 99.9 %, -150 mesh). Milling of NaH and Al without dopant was also performed for comparison.

Table 2.1: Milling parameters for the synthesis of $NaAlH_4$ from NaH + Al with dopant

Milling type	*Starting mixture*	*Balls number (diam.)*	*BPR*[a]	*p (H_2) (bar)*	*Milling duration (h)*	*Rotational speed (rpm)*
1	NaH + Al + 4 mol% $TiCl_3$	7 (15 mm)	50/1	110	4+6	500
2	NaH + Al + 4 mol% $TiCl_3$	37 (10 mm)	50/1	110	5	550
3	NaH + Al + 4 mol% $TiCl_3$	37 (10 mm)	50/1	10, 50, 100, 140	5	550
4	NaH + Al + 2, 4 or 6 mol% $TiCl_3$	37 (10 mm)	50/1	100	6 (2mol%) 5 (4 + 6mol%)	550
5	NaH + Al + 4 mol% $TiCl_3$, Ti, $ScCl_3$ or $CeCl_3$	37 (10 mm)	50/1	100	5	550

[a] Ball-to-Powder Ratio

For the preparation of $Ca(BH_4)_2$, the starting materials were CaH_2 (99.99%, Sigma Aldrich) and CaB_6 (99.5%, Alfa Aesar) with different additives (**Table 2.** 2). To perform the synthesis of $Ca(BH_4)_2$ through the reverse reaction (1.11), CaH_2 and CaB_6 were mixed with 2:1 molar ratio and milled with 140 bar starting hydrogen pressure. Starting from the decomposition products has the advantage to avoid the formation of by-products. An addition of 4 or 8 mol% of additives was given to the starting mixture (2 CaH_2 + CaB_6 + 0.04 or 0.08 additives) to promote the reaction kinetics. Two compositions were used with two different additives $TiCl_3$ (99.999%, Sigma Aldrich) and TiF_3 (Sigma Aldrich).

Table 2. 2: Milling parameters for the synthesis $Ca(BH_4)_2$ from CaH_2 + CaB_6 with additives

Milling type	*Starting mixture*	*Balls number (diam.)*	*BPR* [a]	*p (H_2) (bar)*	*Milling duration (h)*	*Rotational speed (rpm)*
1	2 CaH_2 + CaB_6 + 0.04 $TiCl_3$	37 (10 mm))	50/1	100	20	550
2	2 CaH_2 + CaB_6 + 0.08 $TiCl_3$	37 (10 mm)	75/1	140	25	550
3	2 CaH_2 + CaB_6 + 0.08 TiF_3	37 (10 mm)	75/1	130[b]	25	550

[a] Ball-to-Powder Ratio
[b] Technical limitation

As-received $NaAlH_4$ (Aldrich, 90%) and $Ca(BH_4)_2$ (Sigma Aldrich), commercially available, were also analysed to serve as reference to the doped samples prepared by ball milling. For most measurement techniques, the as-received $Ca(BH_4)_2$ (Sigma Aldrich, dry powder) was composed mainly of the β-phase with presence of a small quantity of the α-phase. For infrared measurements performed at the University of Geneva, the $Ca(BH_4)_2$ product analysed was composed mainly of the α-phase (from drying $Ca(BH_4)_2$.2THF, Sigma Aldrich).

2.2 Structural characterisation

2.2.1 X-ray Diffraction (XRD)

Three types of standard laboratory XRD measurements were performed on the samples. In addition, in-situ measurements with synchrotron radiation were done for the analysis of some $NaAlH_4$ samples.

To allow a direct comparison of the XRD patterns measured at different wavelengths, the diffracted intensities are plotted as a function of the wave vector Q rather than the angle 2θ. The wave vector is defined by:

$$Q = \frac{2\pi}{d} \quad \textbf{(2.1)}$$

where d is the interplanar distance, which is obtained from the Bragg law:

$$\lambda = 2d \cdot \sin\theta \quad \textbf{(2.2)}$$

where λ is the wavelength of the X-ray radiation and θ the diffraction angle.

Room temperature (ex-situ) measurements for $NaAlH_4$ based samples were performed on a Philipps Expert diffractometer with Co Kα radiation (1.79 Å). The samples were placed on a flat holder covered with a Kapton® foil to protect them from air contamination. For $Ca(BH_4)_2$-based samples, room temperature XRD analysis was performed using a STOE STADI MP diffractometer with Mo Kα radiation (0.71 Å). The measurements were done in transmission using a Debye-Scherrer geometry and a Position Sensitive Detector (PSD). In-situ measurements during heating were performed on a Bruker D8 Advance apparatus with Cu Kα radiation (1.54 Å). For transmission and in-situ XRD measurements, a glass capillary (0.7 mm external diameter) was filled with the powder and then sealed by melting the end of the capillary. For in-situ analyses, the capillary was installed inside a MRI (Materials Research Instruments) high temperature capillary furnace.

Synchrotron measurements were performed in Lund (Sweden) in cooperation with T.R. Jensen's group (University of Aarhus, Denmark). Powder was put in a sapphire sample holder. The synchrotron radiation was 1.09801 Å. Each scan was measured for 30 s and was acquired in-situ while changing the temperature or pressure.

Rietveld refinements of the XRD patterns were done using the MAUD software (Material Analysis Using Diffraction) developed by L. Lutterotti [165]. The Rietveld refinement method allows to determine the composition of the sample as well as the structural properties (cell parameters, crystallite size, strain) of the different phases. In MAUD, the average crystallite size and strain values come from an automatic analysis of the specimen peak broadening (across the various hkl peaks) described by analytical Pseudo-Voigt lines. The errors for the determination of the cell parameters or crystallite size are given by the goodness of fit of the refinement. For all $NaAlH_4$ samples, R_w was about 13% so that the values obtained can be compared for all the as-milled samples.

2.2.2 Microscopy and spectroscopy

Scanning electron microscopy (SEM) was performed with a FEG-SEM LEO 1530 Gemini apparatus (20 kV). The samples were prepared in the glove box and transferred to the SEM chamber without contact with air using a transfer unit from Kammrath & Weiss. Enery-dispersive X-ray spectroscopy (EDX) was performed using a Bruker EDX system with a XFlash detector. The transfer of samples in the special transfer unit is crucial for a good analysis of the complex hydride powders by SEM because of their sensitivity to air and water. A short exposition to air (some seconds) between the removal of the sample from the Ar container and the installation in the SEM chamber can lead to a dramatic change of the powder morphology (**Figure 2.2**). After contact with air, the powder is composed of porous particles, formed by the release of H_2 produced by the reaction of $NaAlH_4$ with water (**Figure 2.2a**). In addition, there is likely the formation of liquid NaOH and the particles show very viscous edges.

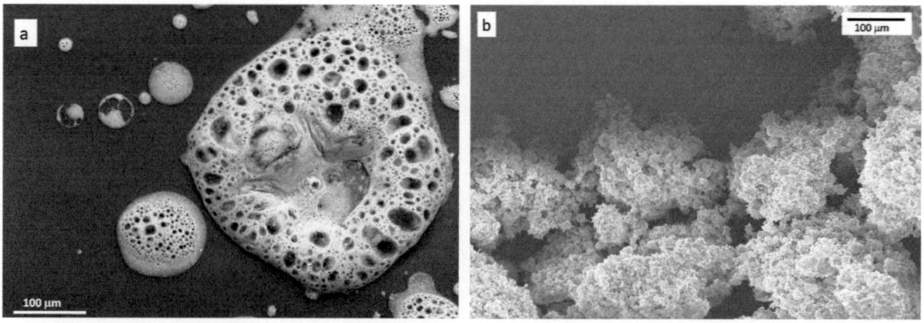

Figure 2.2: SEM pictures for as-milled $NaAlH_4$ with 4 mol% $TiCl_3$: *a)* after short exposure to air for installation in the SEM chamber (BSD detector) and *b)* using the transfer unit between the glove box and the SEM chamber (SE detector).

Transmission electron microscopy (TEM) was performed at ICMSE in Seville (Spain) in cooperation with A. Fernandez and E. Deprez. The microscope is equipped with a parallel detection Electron Energy Loss Spectrometer (EELS). The EELS technique is an analytical technique that measures the change in kinetic energy of electrons after they have interacted with a specimen. Coupled with TEM, EELS can give structural and chemical information about a solid, down to the atomic level. The incident electrons have a specific energy but after transmission through a thin specimen, they emerge with an energy distribution that is characteristic of the material analysed. Energy loss occurs because of the inelastic scattering by both the inner- and outer-shell electrons present in the material. The inner-shell excitation gives rise to ionisation edges, which allow the identification of the elements present. In addition, the ionisation edges are modulated by the fine structure and can be analysed in terms of electronic and atomic structure of the specimen, in particular the shape of the electron loss peak depends on the type of compound present [166].

The TEM analyses were performed on a TEM Phillips CM200 apparatus. In the glove box, a small amount of the powder was diluted in toluene. A droplet of the mixture toluene/powder was then deposited on a copper grid for observation. A vacuum gate valve allowed the sample to be withdrawn and isolated in the chamber during transfer. This procedure reduces the contamination by air during transfer. Energy-Dispersive X-ray (EDX) analyses were carried out at 200 kV. The microscope is equipped with an Electron Energy Loss Spectrometer (EELS) from Gatan (766-2 kV). The EELS spectra were normalised in intensity so that the ratio of intensities measured before and after the edge were equal to one for all measurements.

2.2.3 X-ray Photoelectron Spectroscopy (XPS)

X-ray photoelectron spectroscopy (XPS) is a quantitative spectroscopic technique that enables to determine the elemental composition and chemical state of the elements present on the surface of a material. XPS spectra are obtained by irradiating a material with a beam of X-rays while simultaneously measuring the kinetic energy and number of electrons that escape from the top 1 to 10 nm of the material. XPS is a surface analysis technique but can be used to investigate below the surface by etching the surface with an ion gun (depth profiling). XPS requires ultra high vacuum (UHV) conditions. The electron binding energy of each of the emitted electrons can be determined by using an equation that is based on the work of Ernest Rutherford:

$$E_B = E_p - (E_K - \varphi) \tag{2.3}$$

where E_B is the binding energy of the electron, E_P is the energy of the incident X-ray photons, E_K is the kinetic energy of the electron measured by the instrument and φ is the work function of the spectrometer.

The binding energy is specific for an element and its electronic configuration. The spectra obtained are composed of different peaks corresponding to each shell of each element. The energy measured for the peaks depends on the chemical state of the element. In addition, the number of detected electrons for each of the characteristic peaks is directly related to the amount of this element within the area (volume) irradiated. It is therefore possible to determine the chemical composition of the surface of the material studied.

XPS analysis was performed on a PHI 5600 CI apparatus (Physical Electronics) using a monochromatic Al Kα source (1486.6 eV) and operating under high vacuum (3 10^{-10} mbar). This technique was used for the analysis of doped-$NaAlH_4$. The powder was placed in a molybdenum sample holder inside the glove box and transferred without contact with air inside the XPS apparatus using a closed chamber. Hydrocarbon contamination was too low in the sample to use the C 1s line as internal calibration, therefore the spectra were calibrated using the Cl $2p_{3/2}$ line at 200.7 eV as reference [95]. High-resolution scans were performed on Ti 2p, Sc 2p and Ce 3d binding energy regions. Analysis of the sample under the surface was done after ion beam sputtering of the surface (Ar^+, 1.5 keV, 2.4 μA) for 2, 7 and 17 min.

2.2.4 Vibrational spectroscopy

Raman and infrared (IR) spectroscopy are based on a similar principle. They are sensitive to the vibrations in the molecules inside the material and use the same region of the electromagnetic spectrum. They differ from the observation and measurement methods performed because IR is an absorption method and Raman a scattering method. Basically, the spectrum obtained is composed of different peaks corresponding to the vibrations of structural groups within a molecule. Each peak arises at a certain wavelength/shift specific to each structural group and vibration mode. An example of vibrational modes usually observed by spectroscopy in a XY_4 molecule is given in **Figure 2.3**. Raman and IR spectroscopy are based on different selection rules for the active vibration modes meaning that the spectra obtained for the same material are different. For IR, the absorption occurs if there is a change in the polarity of the molecule. The Raman effect is much weaker. Therefore, laser light is used for the measurement. The light is scattered with a wavelength different from the incident wavelength (the difference is called Raman shift) if there is a modification of the polarisability of the molecule. This change of wavelength results from the change of the molecular vibrational motions.

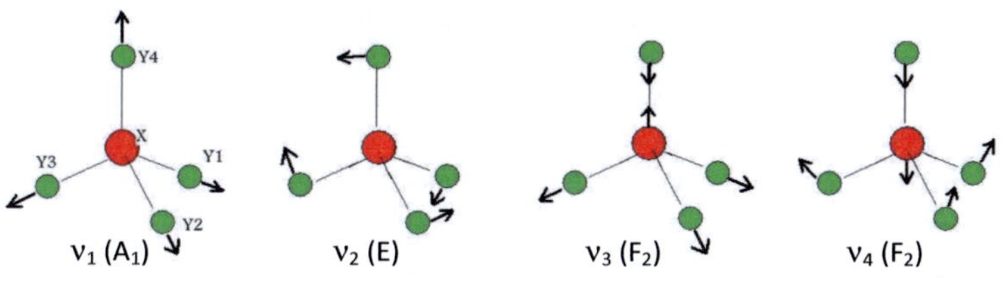

Figure 2.3: Normal vibrational modes in XY_4 tetrahedral molecule [167]. The vibrations ν_1 and ν_3 are related to the stretching modes and ν_2 and ν_4 to the bending modes in the XY_4 molecule.

The advantage of these techniques compared to XRD is that they are less sensitive to the long-range order of the material structure. This means that it is possible to detect compounds highly disordered or even liquid. This is particularly interesting for the analysis of complex hydrides that have very complicated structures, sometimes difficult to detect by XRD, and have low scattering intensities. IR and Raman spectroscopy are complementary because the selection rules are different; it is common that one mode, which is not Raman active, is IR active and vice versa. For borohydrides, the IR or Raman spectra usually show the presence of the stretching and bending modes for the $[BH_4]^-$ units given schematically in **Figure 2.3**.

Raman spectroscopy was performed at room temperature using a Renishaw Ramascope 2000 (Renishaw plc, Gloucestershire, U.K.) with a spectral resolution of 1 cm^{-1}. The 633 nm line of the HeNe laser was focused on the sample through the magnifying objective (50x) of the microscope. Around 10 mg of sample was placed in a Linkam THMS600 cell (Linkam Scientific Instruments, U.K.) directly adapted to the microscope of the instrument. The cell allows the use of controlled atmosphere and temperature. The glass window of the cell is 0.1 mm thick. The samples were placed under a small Ar flow to avoid contamination during measurement.

Infrared (IR) spectroscopy measurements were performed in cooperation with V. D'Anna and H. Hagemann at the University of Geneva. For the measurements, all samples were handled in a nitrogen filled glove box. The samples were introduced in a Specac Golden Gate ATR cell. The IR spectra were recorded at room temperature using a BIO-RAD EXCALIBUR spectrometer with 1 cm^{-1} resolution.

2.2.5 BET (Brunauer-Emmett-Teller) analysis

The BET technique was developed by Brunauer, Emmett and Teller in the early 1940's to determine the specific surface of porous materials based on the physical adsorption of gases on the external and internal surfaces [168, 169].

A material surrounded by a certain gas in equilibrium at a certain temperature T and relative vapour pressure p/p_0, physically absorbs a certain amount of gas (adsorbate). The amount of adsorbed gas depends on the relative vapour pressure and is proportional to the specific surface (external and internal) of a material. The relation between the vapour pressure and the amount of gas absorbed is given by an adsorption isotherm. Assuming that the gas molecules physically adsorb on a solid in infinite layers, that there is no interaction between the layers and that the Langmuir theory applies for each layer, the following BET equation can be established:

$$\frac{1}{v[(p_0/p)-1]} = \frac{c-1}{v_m c}\left(\frac{p}{p_0}\right) + \frac{1}{v_m c} \tag{2.4}$$

where v is the adsorbed gas quantity, v_m the monolayer gas quantity and c the BET constant.

From the adsorption isotherm v vs. (p/p_0) (**Figure 2.4a**), it is possible to draw the BET plot $1/[v(p_0/p)-1]$ as a function of (p/p_0), from which the slope A and the intercept I can be derived for $0.05<(p/p_0)<0.3$ (**Figure 2.4b**).

These values are used to determine the quantity of gas in a monolayer v_m and the BET constant c:

$$v_m = \frac{1}{A+I} \tag{2.5}$$

$$c = 1 + \frac{A}{I} \tag{2.6}$$

The specific surface area (BET surface) is proportional to v_m and is given by:

$$S_{BET} = \frac{v_m \cdot N \cdot s}{V_M} \tag{2.7}$$

where N is the Avogadro number (6.022 10^{23} mol^{-1}), s the cross section of the adsorbate and V_M the molar volume of the adsorbate.

The BET analysis was performed using a Belsorb Max apparatus from BEL Japan Inc. with N_2 adsorbate gas. The molar volume of N_2 is 13.54 10^{-3} $m^3\,g^{-1}$ and the cross section 16.2 Å [169]. The adsorption is performed at liquid N_2 temperature -196°C (77 K). The as-milled powders were loaded in glass capillaries and the dead volume was calibrated by several cycles of evacuation and N_2 filling at room temperature prior to the measurement. After the first determination of the surface for milled samples during adsorption and desorption of N_2 at -196°C, the capillaries were heated to 300°C under vacuum to decompose the sample and then another cycle of N_2 adsorption and desorption at -196°C was performed.

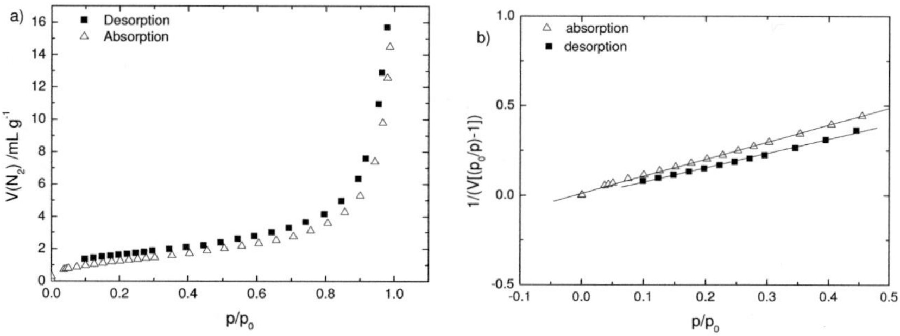

Figure 2.4: a) adsorption isotherm for N_2 at -196°C (77K) and b) BET plot derived from a) for the as-milled sample $NaAlH_4$ + 4mol% $TiCl_3$

2.3 Thermal analysis

Considering the final application as hydrogen storage materials, the thermal analysis is focussed on the study of the reactions during hydrogen uptake/release. These reactions involve gas-solid interactions and the pressure is a very important parameter. The different methods used give complementary information about the reactions taking place.

2.3.1 Differential scanning calorimetry (DSC)

Differential scanning calorimetry analyses the heat (ΔQ) exchanged during heating/cooling of a sample. The sample and a reference are heated at the same rate. Enthalpy or heat capacity changes in the sample cause a difference in its temperature relative to the reference. The apparatus measures the heat flow between the sample and the reference produced to compensate this difference. This provides qualitative and quantitative information about the physical or chemical changes in the sample that involve endothermic or exothermic processes.

DSC traces can be used directly to analyse the sequence of reactions taking place during heating/cooling with or without hydrogen atmosphere. In addition, the onset temperatures of the endothermic or exothermic peaks measured under hydrogen pressure depend on the van't Hoff equation (2.8). Here *p* is the hydrogen partial pressure, R the gas constant (8.314 J K^{-1} mol^{-1}), *T* the absolute temperature, ΔH the enthalpy of H_2 ab- or desorption and ΔS the entropy of the reaction.

$$\ln p = -\frac{\Delta H_{abs/des}}{RT_{abs/des}} + \frac{\Delta S}{R} \quad \textbf{(2.8)}$$

This means that the reaction involving hydrogen (ab-/desorption) occurs at different temperatures for different H_2 pressures. In a first approximation, the formation/ decomposition of a hydride can be regarded as a simple exothermic/endothermic reaction, similar to, e.g. melting or a structural transformation. The distinction between such processes can be difficult from the DSC traces. As an example, the DSC traces of Ti-doped Na_3AlH_6 under 120 bar H_2 are given in **Figure 2.5**. This figure shows peaks attributed to a structural transformation of Na_3AlH_6 (α- to β-phase) at 250°C (T_{eq}^{struc}) and the ab- and desorption of hydrogen at 277°C (T_{eq}^{abs}) and 288°C (T_{eq}^{des}), respectively. Assuming quasi-equilibrium, the Gibbs free energy ΔG for this process is zero, i.e.

$$0 = \Delta G = \Delta H - T_{eq}\Delta S \quad \textbf{(2.9)}$$

where ΔH and ΔS are the corresponding enthalpy and entropy of the reaction, respectively, and T_{eq} is the equilibrium temperature, for example, the melting temperature. T_{eq} is the temperature at which a reaction will start considering only thermodynamics (in most cases an extrapolation). For DSC, such temperature is assumed to be the onset temperature of the peak. This temperature represents the starting of the reaction and the point where ΔG becomes different from zero (eq. (2.9)).

The described situation is fulfilled in most phase transformations (melting and crystal structure transformations) as shown in **Figure 2.5**. During the DSC measurement, the released heat ΔQ is given by:

$$\Delta Q = \int \frac{dQ}{dT} dT - Background \quad \textbf{(2.10)}$$

where dQ/dT is the heat flow measured for maintaining the same temperature for the sample and the reference.

So assuming equilibrium conditions and that the reaction starts at T_{eq} considering only thermodynamics:

$$\Delta Q = T_{eq} \Delta S \quad \textbf{(2.11)}$$

Thus ΔQ is a measure of the enthalpy of formation ΔH according to equation (2.9). This can be directly applied to the structure transformation for which the peak width is less than two degrees, and no difference between T_{eq} for heating and cooling is found. The situation is different for hydrogen sorption. The corresponding peak width is approximately ten degrees, and the extrapolated equilibrium temperatures for heating and cooling differ. The reason for these differences is the relatively slow kinetics for sorption processes compared to structural transformations. Therefore, simple equilibrium conditions cannot be assumed and thus the measured heat exchange ΔQ does not resemble ΔH.

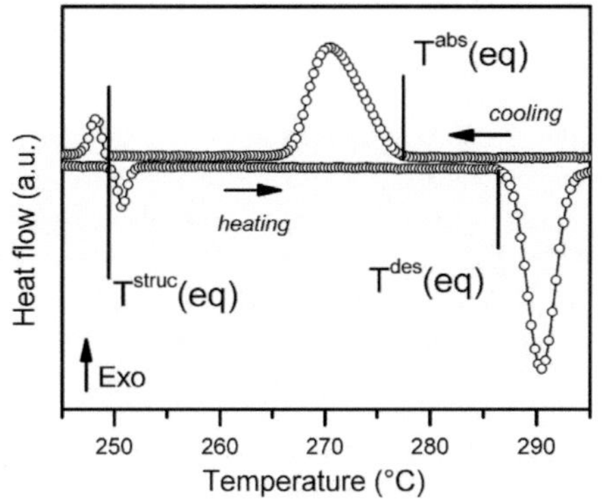

Figure 2.5: DSC traces for $TiCl_3$ doped-Na_3AlH_6 at 5 K min^{-1} and 120 bar H_2

Nevertheless, the DSC measurements can be used to approach the thermodynamic properties of the reactions. The DSC signal is used to determine the temperature at which the ab-/desorption starts. However, in many hydride systems, the onset (at which a signal is visible) is determined by kinetics instead of by thermodynamics. In general, the experimental values of T_{eq} (absorption T_{eq}^{abs}, desorption T_{eq}^{des}) obtained by an extrapolation of the DSC signal to zero, give the range in which the true T_{eq} lies (Figure 2.5):

$$T_{eq}^{abs} < T_{eq} < T_{eq}^{des} \quad \textbf{(2.12)}$$

From isobaric DSC experiments, it is possible to extract ΔH from the van't Hoff plots using T_{eq}^{i} measured under various hydrogen pressures p. Moreover, by including hysteresis in the determination of T_{eq}, the true ΔH value is expected to be between the slopes for absorption and desorption in the van't Hoff plot:

$$slope^{abs} < \Delta H < slope^{des} \quad \textbf{(2.13)}$$

DSC measurements were performed using a Netzsch apparatus (DSC 204 HP Phoenix, heat flow apparatus), which is placed inside a dedicated glove box under Ar. The heating rate was fixed to 5 K min^{-1} for all the measurements. Different pressures between 3 and 140 bar H_2 or 3 bar Ar were used for these measurements. The set pressure was kept constant during all measurements using a pressure controller that compensated the release/uptake of hydrogen by the sample (dynamic mode). Some milligrams of the sample were placed in an Al_2O_3 crucible closed with a lid containing a hole for the gas exchange.

The reproducibility of the measurements was very good. Similar DSC traces were obtained for the samples measured several times or from different milling batches. The onset temperatures of the peaks were determined from the derivative of the DSC traces with some degrees Celsius of error.

For samples related to $Ca(BH_4)_2$, the ΔQ measured was very small and a large variation of the baseline with temperature was observed. Several parameters may affect the measurement. In this case, the samples were supposed to have low thermal conductivity, high thermal resistance and large heat capacity [170], likely related to the presence of boride compounds. To show only the ΔQ related to the sample, the background was removed from the DSC traces by measuring the traces for an empty crucible under identical conditions (zero line).

2.3.2 Thermogravimetry (TG)

The thermogravimetry measurements were performed on a Hiden Isochema IGA (Intelligent gravimetric analysis, type IGA-001) apparatus. This system is composed of a micro-balance that measures the mass variation of the sample during heating/cooling with 0.1 μg resolution. This system was used to analyse the release of hydrogen (decrease of the sample mass) during heating in Ar atmosphere. The results gave the hydrogen capacity of the sample.

The powder (20-25 mg) was placed in a stainless steel basket that is attached to the balance via a hooked wire. The sample was transferred from the glove box to the balance under Ar by using a "dry loading" system developed by Hiden. Prior to connecting the sample to the balance, the dry loader was evacuated and refilled with Ar (35 cycles) to reduce as much as possible the oxygen and water content. Once the sample hung to the hook and the reactor closed, the temperature was increased to the desired temperature with an apparent heating rate of 5 K min^{-1} under 1 bar Ar atmosphere.

For thermogravimetric measurements, some errors can arise from the so-called buoyancy effect.

The buoyancy forces are opposite to gravity for a specimen placed in a fluid (here Ar gas). When the temperature increases, the forces are changed by the change of pressure. An apparent weight change can be detected if only the sample side of the balance is heated and not the counterweight. This was the case here and we observed a small apparent weight increase at the beginning of the heating stage below 50°C. Afterwards, the measured weight was constant before decomposition of the sample. The TG curves shown in this work are plotted from 50°C and no buoyancy effect can be seen. The reproducibility of the measurement was affected by the quality of the vacuum obtained prior of measurement. Some oxidation of the sample can occur and increases the weight of the sample.

2.3.3 Pressure-Composition Isotherms (PCI)

The measurement of several pressure-composition-isotherms (PCI or pcT for pressure-composition-temperature diagram) is the dedicated method for the determination of the thermodynamics properties of a reaction. This is in principle an equilibrium method that gives the phase diagram of the reaction (**Figure 2.6**). The hydrogen absorption/desorption reaction is characterised by a plateau pressure (miscibility gap) depending on the temperature and the thermodynamic properties (enthalpy and entropy) of a compound (see §1.1.2.2). In experimental conditions, a hysteresis exists between absorption and desorption because the kinetics of the reaction hinder to fully reach the equilibrium conditions.

The PCI were here determined using a Sieverts' type apparatus based on a volumetric method. This method relies on a volume perfectly known for the sample holder and for the reservoir (where the gas will expand) used. The hydrogen capacity is calculated from the pressure variations measured for a constant temperature using the real gas law. For a given temperature, the sample undergoes increasing (absorption) or decreasing (desorption) pressure steps and the variations of the pressure are recorded until equilibrium is reached (constant pressure). Under practical conditions, it is very difficult to determine the equilibrium point. So the measurement time is set to a sufficiently long time to approach this state. Each pressure step gives a point in the pressure-composition diagram (quantity of H_2 release/uptake vs. final "equilibrium" pressure). The pressure steps are done until the full capacity (in ab- or desorption) is reached. The enthalpy and the entropy of the reaction are obtained by constructing a van't Hoff plot with $\ln p$ as a function of $1/T$ (**Figure 2.6**). According to the van't Hoff equation (eq. (2.8)), the slope gives the enthalpy of reaction and the intercept with the Y-axis gives the entropy of the reaction. For the construction of a van't Hoff plot, the determination of several isotherms is necessary. The enthalpy and entropy values obtained from the van't Hoff plot for absorption and desorption (because of the hysteresis in plateau pressure) give the lower and upper bounds of the true values.

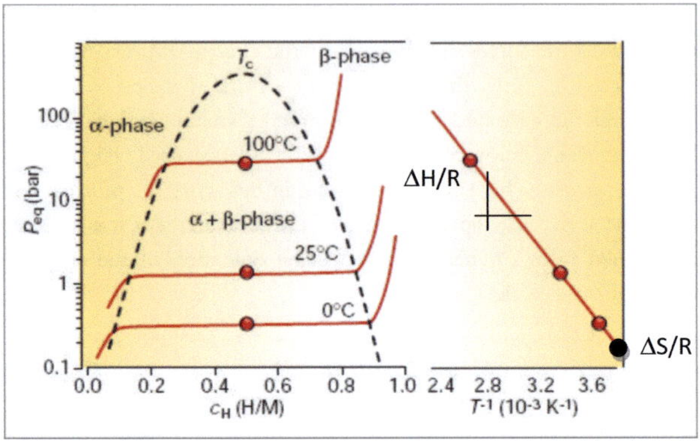

Figure 2.6: Example of PCI curves and van't Hoff plot for $LaNi_5$ (from [7])

The Sieverts apparatus used was a fully automatic Hy-Energy PCTPro2000 system. A large quantity of powder (ca. 200 mg) was loaded in a stainless steel cylinder, which was placed in a stainless steel sample holder (max. 200 bar/400°C). The temperature of the sample was controlled by an external jacket heater placed around the sample holder. The volume of the reservoir was chosen to adjust the final pressure and the reservoir was kept at a constant temperature of 31 ± 0.5°C, which was maintained by an internal heater. To limit the errors originating from the difference of temperature between the sample and the reservoir, the free volume in the sample holder was set as small as possible using stainless steel spacers. The volume of the holder was calibrated before the measurement at the desired temperature to take into account the dilatation of the different components.

This apparatus was also used to analyse the kinetics of hydrogen ab-/desorption isothermally following the pressure changes with time. The quantity of hydrogen uptake/release was calculated using the real gas law. With a Sieverts' apparatus, the measurements are always performed with a H_2 back pressure. For desorption, the reservoir was initially evacuated before opening the sample holder (under Ar or H_2). Then, the measurement is performed in a closed volume with the H_2 back-pressure depending on the initial H_2 pressure in the sample holder, the volume of the reservoir and the quantity of H_2 released from the sample. The largest reservoir volume possible (ca. 1 L) was chosen during desorption measurements to avoid an increase of the H_2 pressure in the reservoir that exceeds the plateau pressure. In these conditions, the back-pressure was about 1 bar H_2. For absorption, the sample holder was kept at a low pressure (after desorption). Before opening, the reservoir was filled with a pressure above the desired hydrogen pressure (calculated to reach the desired pressure when it is connected to the sample holder). The volume of the reservoir (170 mL) was chosen to limit the H_2 consumption during filling and to avoid a large pressure variation when H_2 is absorbed (here 1-2 bar) by the sample that would change the measurement conditions.

2.3.4 Thermal Desorption Spectroscopy (TDS) with mass spectrometry

Thermal desorption spectroscopy (TDS) is a technique that allows to measure the evolved gas from a sample during heating. It is not possible to evaluate the quantity of H_2 released from the spectra obtained without a careful calibration of the gas release from known compounds that was not available for the apparatus used.

The samples (10 mg) were filled into a stainless steel reactor under Ar atmosphere with a dew point below -70 °C. Prior to each TDS measurement, the UHV chamber was evacuated to reach a base pressure below 10^{-8} mbar. The desorbed products were detected with a quadrupole mass spectrometer MS (Balzers QMG 064) mounted in line of sight of the reactor. A mass-to-charge ratio m/e =2 was used for the detection of H_2 and m/e = 26 for the detection of the diborane (B_2H_6) by-product (Pfeiffer Vacuum mass spectra library). During the TDS measurements, no impurity gasses (e.g. H_2O, CH_4, CO, CO_2) were observed above 10^{-8} mbar, except for CaH_2 that releases a larger quantity of H_2O, as commonly observed for this compound [171].

3. $NaAlH_4$: a model system

Study of the catalytic mechanism

Sodium alanate ($NaAlH_4$) is the first complex hydride that has been studied for solid state hydrogen storage applications. The work by Bogdanovic and Schwickardi in 1997 [18] on the improvement of the kinetics of hydrogen release/uptake by adding $TiCl_3$ has triggered an enormous interest in the study of complex hydrides like alanates or borohydrides. The hydrogen sorption mechanisms are well established and the kinetics properties have already been dramatically improved for $NaAlH_4$. Nevertheless, there are still open questions on the detailed catalytic mechanism, i.e. on the actual catalysis reaction(s) and the role of the dopant. Despite a large diversity of analytical techniques used to study in particular the chemical state of Ti, the exact role of the dopant $TiCl_3$ is not known. The role of Ce- or Sc-species is even less known although $CeCl_3$ or $ScCl_3$ are claimed to be more efficient than $TiCl_3$.

The aim of this chapter is in first to establish the synthesis process for doped $NaAlH_4$ by reactive ball milling. The information gained from these investigations will be afterwards applied to the synthesis of other complex hydrides. The influence of different dopants on the thermodynamics and kinetics of $NaAlH_4$ is systematically described in a second part. Finally, the last section is focussed on the analysis of the catalytic mechanism, in particular by comparing the effect of different dopants and in the light of the results obtained previously (**Annexe C1**).

3.1 Synthesis by reactive ball milling

3.1.1 Influence of the milling parameters on the synthesis process

Reactive ball milling has been successfully applied to the synthesis of different classes of hydrides (see §2.1). This technique has the advantage that the catalyst/dopant can be added directly with the starting reactants so that synthesis and doping are performed in a one-step process. For $NaAlH_4$, it has been shown that better sorption properties can be obtained by this technique [77]. For the preparation of $NaAlH_4$, NaH and Al are milled together in H_2 atmosphere.

In this first part, the effect of different milling parameters is analysed when $TiCl_3$ is used as dopant. The initial milling parameters (milling type 1) are given in **Table 2.1**. The hydrogen pressure was set to 110 bar and the milling time was 4 h. The evolution of the pressure and temperature during milling is given in **Figure 3.1a**. A pressure rise up to 115 bar is observed initially, which is related to a temperature increase in the vial because of the mechanical activation, i.e. friction between balls and vial walls. After 3 h milling, pressure and temperature stay almost constant indicating that no further reaction between the powders and H_2 occurs inside the vial. This is more clearly evidenced by the absence of variation of the H_2 content in the vial (**Figure 3.1b**). The quantity of H_2 is calculated from the pressure and temperature recorded during milling and using the ideal gas law. Similar observations are made by restarting the milling for six additional hours and no reaction is again observed for this longer process time (total 10 h).

The comparison of the X-ray diffraction (XRD) patterns of the starting mixture and the as-milled powder (**Figure 3.2a** and **b**, respectively) shows that the starting products NaH and Al are still present after 10 h milling; the peaks are only broadened due to grain refinement. This result confirms that no reaction occurred during milling.

For mechano-chemical synthesis, the reaction occurring for a defined system is characterised by an ignition temperature (T_{ig}) that decreases during the milling because of the decreasing crystallite size [172, 173]. The reaction starts after an incubation time, i.e. the time necessary for the local temperature inside the vial to reach T_{ig}. For milling type 1, the absence of a reaction indicates that the local temperature is not high enough to reach T_{ig}. The local temperature can be risen up by increasing the energy input and the power absorbed by the material. The power consumption P by the powder can be evaluated using equation (3.1) considering the collision model [174]. In this equation, m_b and n_b are the weight and number of balls, w_p and R_p the speed and radius of the planetary mill, P^* a dimensional coefficient depending on the geometry of the mill and the elasticity of the collisions.

$$P = P^* \frac{1}{2} m_b w_p^3 R_p^2 n_b \qquad \textbf{(3.1)}$$

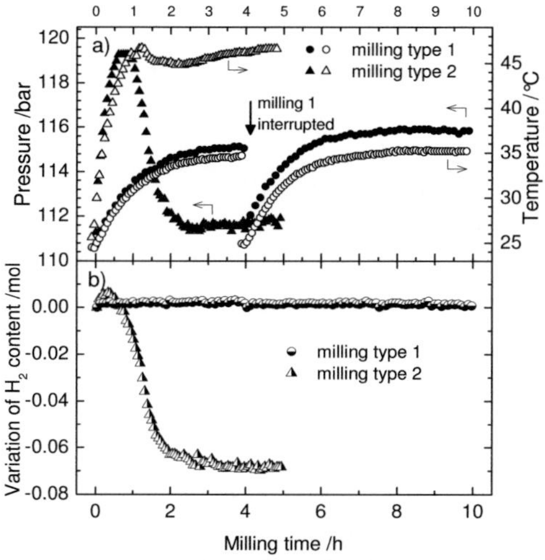

Figure 3.1: a) variation of pressure (closed symbols) and temperature (open symbols) and b) of the H_2 content (half filled symbols) in the vial for NaH + Al + 4mol% $TiCl_3$ milled with different parameters (Table 2.1). Milling type 1: low energy input, milling type 2: high energy input.

An increase of the power input can be achieved by increasing the number of balls or the rotational speed. The milling parameters were therefore varied (milling type 2, see **Table 2.1**). The increase of the local temperature compared to milling type 1 is reflected by the increase of the temperature measured in the lid (**Figure 3.1a**). After one hour, the local temperature reaches T_{ig} as suggested by the subsequent decrease of the pressure indicating the absorption of H_2 and the formation of $NaAlH_4$. The reaction is finished after only two hours milling and the pressure remains constant. Two maxima in temperature are noticeable, simultaneous to the pressure drop, that could be related to the two exothermic steps of the formation of $NaAlH_4$ [49] (see §1.2.1).

The pressure drop obtained during milling can be used as a tool to evaluate the reaction in-situ. After cooling the vial to room temperature, a total pressure drop Δp of 9.2 bar is calculated using the ideal gas law (**Table 3.1**, p. 54). Based on a complete reaction between NaH and Al with H_2 to form $NaAlH_4$, a Δp of 10.6 bar H_2 is expected. This information seems to indicate that the reaction is not complete after 5 h milling despite no pressure variation is observed after 3 h milling. XRD analysis performed on the as-milled sample (**Figure 3.2c**) confirms the formation of $NaAlH_4$ and indicates the presence of NaCl and un-reacted Al. These compounds are the by-products of the reduction of $TiCl_3$ to Ti metal occurring during milling [95] (equation (3.2)).

$$3NaAlH_4 + TiCl_3 \rightarrow 3NaCl + 3Al + Ti + 6H_2 \qquad \textbf{(3.2)}$$

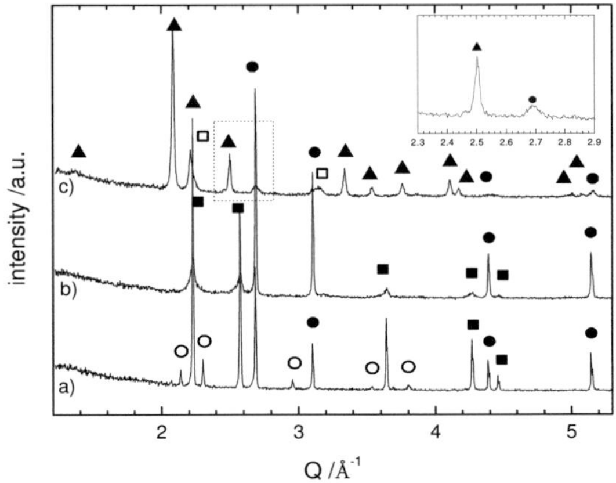

Figure 3.2: XRD patterns for a) the starting powder mixture, and for the powders milled with b) milling type 1 (low energy input) and c) milling type 2 (high energy input). Inset: enlargement of pattern c) (square). The phases present are: $NaAlH_4$ (▲), Al (●), NaCl (□), NaH (■) and $TiCl_3$ (O).

The deviation of the measured pressure drop from the theoretically expected value is thus explained by the concomitant reduction of $TiCl_3$ to Ti^0 (metal) leading to the formation of NaCl instead of $NaAlH_4$. Considering this reaction, the Δp calculated is then 9.4 bar, being close to the experiment (**Table 3.1**, p .54), and confirming the completion of the synthesis reaction after 5 h milling.

On the other hand, it is not possible to detect any pure Ti phase from the XRD patterns. Ti may have formed an Al-Ti alloy as already reported [106, 175], which may be related to the very broad and asymmetric peak of Al at Q=2.7 Å^{-1} (inset in **Figure 3.2**) towards the high value of Q (angle) as described by Haiduc et al. [106].

Rietveld refinement asserts that NaH and Al have entirely reacted to form $NaAlH_4$. The data obtained from the Rietveld refinement are summarised in **Table 3.2** (p. 57), together with the expected weight fractions of the different phases (considering the full decomposition of $TiCl_3$ and the complete formation of $NaAlH_4$). The amounts of $NaAlH_4$ and NaCl obtained after milling are similar to what is expected. The fraction of Al phase in the milled sample corresponds to the fraction expected for this phase and for Ti, in agreement with the formation of Al-Ti compounds during milling. The cell parameters determined by the refinement for the Al phase do not change significantly compared to the initial Al phase disclosing the substitution of Ti in the Al structure. A new Al-Ti compound is likely to be formed appearing at higher Q, but may be overshadowed by the broadening of the Al peak. Additionally, the Rietveld refinement reveals that the crystallite sizes for all phases are really small around 25 nm for Al and NaCl and 53 nm for $NaAlH_4$, confirming the high efficiency of the ball milling technique to obtain nanocrystalline powders. The calculated strains are very small (less than 1%).

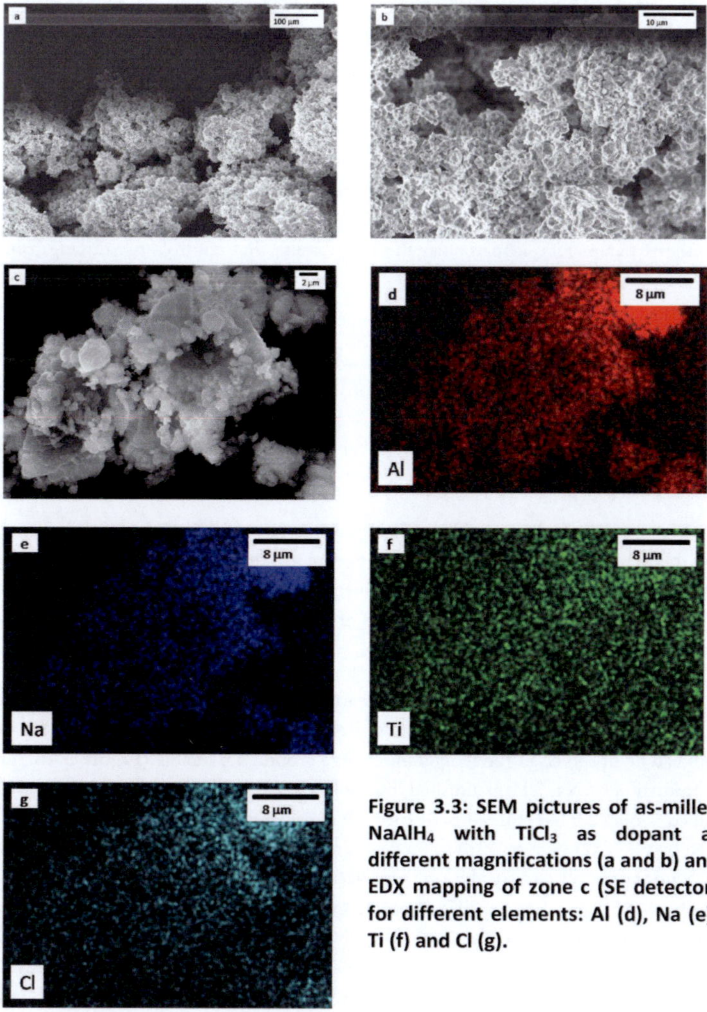

Figure 3.3: SEM pictures of as-milled $NaAlH_4$ with $TiCl_3$ as dopant at different magnifications (a and b) and EDX mapping of zone c (SE detector) for different elements: Al (d), Na (e), Ti (f) and Cl (g).

The morphology of the as-milled sample has been analysed by SEM. The micrographs taken at different magnifications are given in **Figure 3.3a and b**. The powder is composed of large aggregates (>100 μm) of small particles (ca. 1 μm). EDX mapping was performed (**Figure 3.3d-g**) on a selected area (**Figure 3.3c**) for one large agglomerate and shows a good distribution of the elements, in particular of the "dopant" Ti that should ensure a high efficiency of the dopant.

3.1.2 Influence of the hydrogen pressure

The XRD and SEM analyses demonstrate that the strategy of milling type 2 (increasing the local temperature to reach T_{ig}) is very efficient to prepare doped-$NaAlH_4$ with a good distribution of the dopant and a nanocrystalline structure. The process may be further enhanced by varying other synthesis parameters, in particular the H_2 pressure that controls the driving force of the reaction (see below).

Different millings of NaH and Al with 4 mol% $TiCl_3$ using different starting pressures (10 to 140 bar H_2) were performed (milling type 3, **Table 2.1**). For the temperature range (45-55°C) of the milling, there is no thermodynamic limitation for the formation of $NaAlH_4$ because the equilibrium pressures are lower than the lowest pressure used for milling, i.e. 10 bar H_2. The equilibrium pressures are 0.1-0.2 and 5.2-7.9 bar H_2 for reactions (1.7) and (1.6), respectively, considering the enthalpies of formation given by Bogdanovic et al. [50]).

The variation of pressure during milling (**Figure 3.4a**) shows the same incubation time of about one hour before hydrogen absorption, independent of the initial pressure. This observation indicates that the hydrogen pressure has no influence on the ignition temperature of the reaction. On the contrary, the starting pressure has an effect on the reaction rate because a faster H_2 absorption is observed for the higher pressures.

For the samples milled with 10 or 50 bar H_2 starting pressure, the rate of H_2 absorption is low and the formation of $NaAlH_4$ is not complete after 5 h. Using 10 bar H_2 pressure, only one weak maximum can be observed in the temperature curve around 4 h milling (**Figure 3.4b**), indicating that the formation of Na_3AlH_6 is finished, but only a small amount of $NaAlH_4$ is expected. The Δp obtained is only 2.5 bar (for 9.4 bar expected) clearly indicating a small reaction yield. This is confirmed by the XRD pattern of the as-milled sample where mainly Na_3AlH_6 and Al are present (**Figure 3.5a**). Using 50 bar H_2, the pressure decreasing rate stabilises after 5 h milling but the drop observed (5.5 bar) is smaller than expected (9.4 bar). Nevertheless, the reaction proceeds in two distinct steps characterised by the well-resolved maxima in the temperature curve and two hydrogen absorption steps meaning that the formation of $NaAlH_4$ from Na_3AlH_6 has started but is not finished. The incomplete formation of $NaAlH_4$ is shown in the XRD pattern (**Figure 3.5b**) by the presence of the intermediate phase Na_3AlH_6.

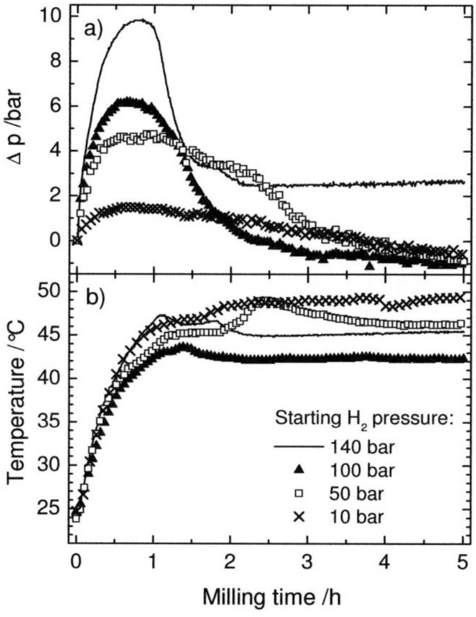

Figure 3.4: variation of a) pressure and b) temperature for NaH + Al + 4 mol% $TiCl_3$ milled with different starting pressures (milling type 3).

The highest total pressure drop (close to the expected Δp of 9.4 bar) and reaction rate are found for the sample milled with 100 bar H_2 starting pressure, where the two absorption steps overlap. No further enhancement is obtained using 140 bar H_2. In both cases, the absorption reaction is finished after ca. 2 h and Δp is comparable. For the samples milled with p>100 bar, the XRD patterns show only the presence of $NaAlH_4$ and the by-products NaCl and Al (**Figure 3.5c and d**).The increase of the reaction rate by increasing the applied hydrogen pressure can be partly explained by an increase of the driving force of the reaction. The rate of hydride formation r can be described by [176]:

$$r = k\left[\frac{(p-p_e)}{C}\right]^a \quad \textbf{(3.3)}$$

where C is the hydrogen concentration in the material, a the order of the reaction, p the applied pressure, p_e the equilibrium plateau pressure and k the rate constant. k can be written as following an Arrhenius-type behaviour as:

$$k = A\exp\left(\frac{-E^a}{RT}\right) \quad \textbf{(3.4)}$$

where A is the pre-exponential factor, E^a the activation energy, R the gas constant and T the absolute temperature.

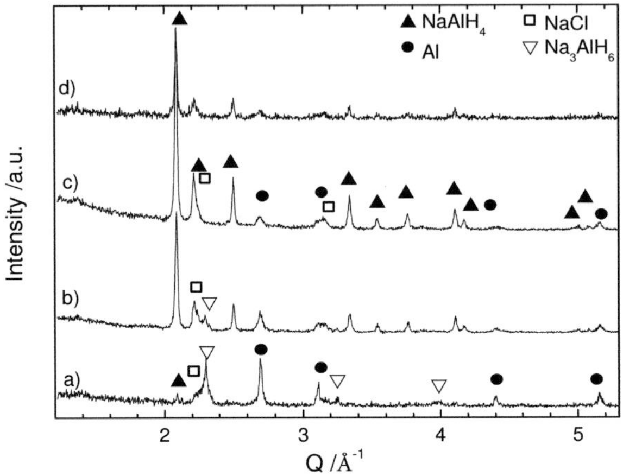

Figure 3.5: XRD patterns for the mixtures NaH + Al with 4mol% $TiCl_3$ milled with different H_2 starting pressures: a) 10 bar, b) 50 bar, c) 100 bar and d) 140 bar H_2.

The effect of the pressure is described by the second term of equation (3.3) $[(p\text{-}p_e)/C]^a$ that is the driving force of the reaction. The equilibrium pressure can be considered the same in each case because the temperature inside the milling vial is similar (**Figure 3.4b**). For the same reason, it is also assumed that k (related to the dynamic properties of the reaction) is of the same order in each case. The main effect of the starting pressure is then clearly to increase the driving force by increasing the $(p\text{-}p_e)$ term. Nevertheless, the increase of the driving force should lead to a faster reaction using 140 bar H_2 as starting pressure that is not observed experimentally. This indicates that this very high pressure has no additional beneficial effect on the milling process.

The use of a reactive atmosphere is known to promote the reduction of the particle size because it limits the cold welding processes [172]. The reduction of the particle size enhances the reaction rate by creating fresh surface areas. The use of a high pressure, in addition to providing a higher reagent content, may also facilitate the particle size decrease and thus the reaction rate. At a certain point, the particle size reaches a final equilibrium value depending on the material and the milling conditions. This may explain why the reaction rate reaches a maximum using 100 bar H_2 and no further enhancement is found using higher pressures. The crystallite sizes determined by Rietveld refinement are close for samples milled with 100 or 140 bar H_2 (53 and 69 nm for $NaAlH_4$, respectively).

3.1.3 Influence of the dopant

3.1.3.1. Effect of dopant content

From the study of the influence of the milling parameters and the H_2 pressure, it appears that the optimised milling parameters for a fast and efficient synthesis of $TiCl_3$-doped $NaAlH_4$ are given by milling type 4, i.e. 100 bar H_2 starting pressure, the use of 37 balls and a rotational speed of 550 rpm (**Table 2.1**). Now, the in-situ monitoring capability is used to analyse the efficiency of the different dopants already during synthesis. A first step is to determine the optimised dopant content. For that purpose, different amounts of $TiCl_3$ were added to NaH and Al in the vial (milling type 4). The pressure and temperature variations (**Figure 3.6**) show the effect of using 2, 4 or 6 mol% $TiCl_3$ on the synthesis process.

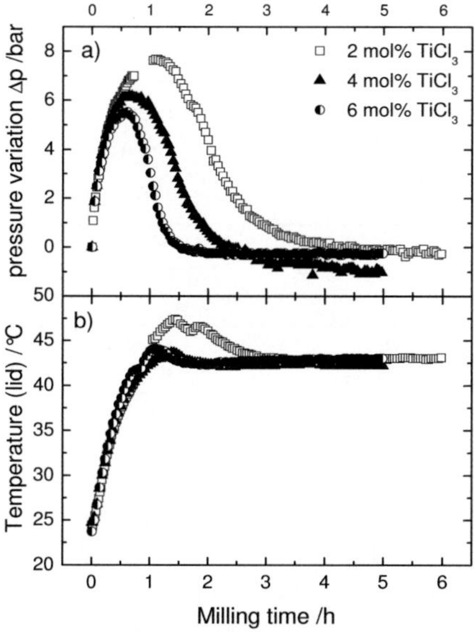

Figure 3.6: Variations of a) pressure and b) temperature during milling of NaH + Al with different quantities of $TiCl_3$ (milling type 4).

As indicated by the shortest milling time needed to observe the pressure drop, the ignition of the reaction (see §3.1.1) occurs faster when using a higher dopant content. The decomposition of $TiCl_3$ may play a role on the local temperature, e.g. by releasing heat. In addition, the rate of the pressure decrease shows that the reaction kinetics are also improved by a higher dopant content. XRD analysis of the as-milled powders (**Figure 3.7**) confirms the formation of $NaAlH_4$ in each case and shows also the presence of NaCl and un-reacted Al as by-products of the reduction of $TiCl_3$ to Ti metal (equation (3.2)).

From the total pressure drop Δp in the vial at room temperature (**Table 3.1**), the reaction appears to be complete only for 4 and 6 mol% dopant (Δp corresponds to the total formation of $NaAlH_4$ and the reduction of $TiCl_3$). On the contrary, for 2 mol% $TiCl_3$, Δp is slightly lower than expected, in agreement with the XRD analysis showing the presence of the remaining intermediate Na_3AlH_6 phase (**Figure 3.7**). The lower calculated Δp for the higher dopant content is related to the decrease of the total hydrogen capacity of the material, because the ratio of non-absorbing compound increases. Therefore, the addition of 4 mol% dopant appears to be a good compromise between fast reaction rate and sufficiently high hydrogen capacity (**Annexe D**).

Table 3.1: Pressure difference Δp between the initial and the final H_2 pressure in the milling vial[a].

Dopant	*$TiCl_3$*			*$ScCl_3$*	*$CeCl_3$*	*Ti*
Content (mol%)	*2*	*4*	*6*	*4*	*4*	*4*
Measured[a]	8.5	9.2	8.1	9.6	8.9	3.3
Calculated[b] /Formation $NaAlH_4$	10.4	10.3	9.9	10.8	9.8	11.0
Calculated[b] /Formation $NaAlH_4$ + reduction dopant [eq. (3.2)]	9.8	9.4	8.3	9.7	8.7	-

[a] pressures taken at room temperature, [b] using ideal gas law

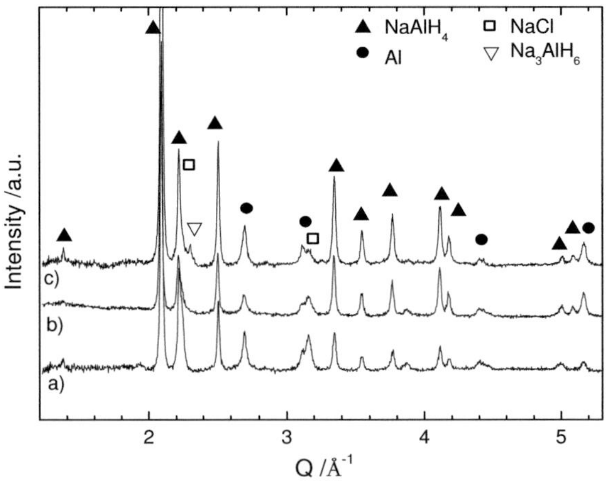

Figure 3.7: XRD patterns for as-milled mixtures NaH+Al with a*)* 6 mol%, *b)* 4 mol% and *c)* 2 mol% $TiCl_3$

3.1.3.2. Study of different dopants

From the results described previously, it is clear that in-situ monitoring the pressure and temperature during milling can give valuable information on the process. Going a step further, the influence of different dopants is studied and the efficiencies are compared with that for $TiCl_3$. The milling parameters defined previously (milling type 5, see **Table 2.1**) were used to prepare $NaAlH_4$ doped with 4 mol% $TiCl_3$, Ti, $ScCl_3$ and $CeCl_3$ (**Figure 3.8**).

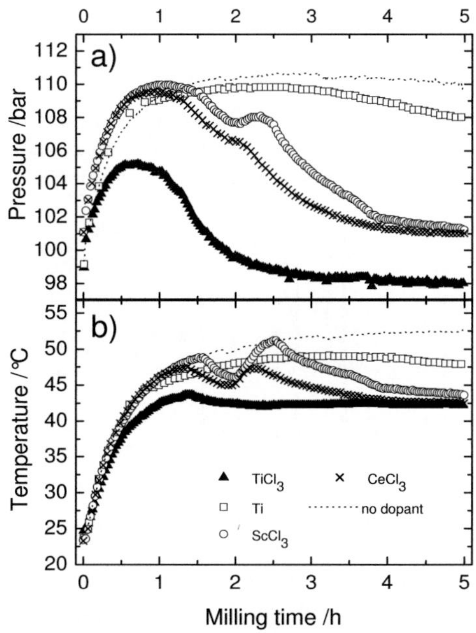

Figure 3.8: Variations of *a)* pressure and *b)* temperature during milling NaH + Al with 4 mol% of different dopants (milling type 5).

The pressure curve (**Figure 3.8a**) related to the milling with Ti shows only a small and slow pressure drop. The efficiency of pure Ti for the synthesis of $NaAlH_4$ is very small, the pressure decrease rate is only slightly higher than in the case of NaH and Al milled without dopant. It is likely that only a small amount of Na_3AlH_6 is formed after 5 h milling in both cases. For $ScCl_3$ and $CeCl_3$, the pressure drop stabilises after 4 h milling and the final Δp after cooling the vial indicates the complete formation of $NaAlH_4$ and the reduction of the dopant through a reaction similar to equation (3.2) for $TiCl_3$ (**Table 3.1**). For the $CeCl_3$ doped sample, the final Δp is slightly larger than the one calculated for a complete reduction of $CeCl_3$, indicative for an incomplete reduction of $CeCl_3$. The milling times for a complete reaction are much longer (3 to 4 h) using

$ScCl_3$ or $CeCl_3$ than using $TiCl_3$ as dopant (**Figure 3.8b**). The reaction rate is lower and the two maxima in temperature can be better resolved than with $TiCl_3$ with two steps of H_2 absorption that can be observed in the pressure variations.

The observations made from the evolutions of pressure and temperature during milling are confirmed by XRD measurements on the as-milled powders (**Figure 3.9**). Patterns for the milling without dopant and using Ti as dopant are found to be very similar: the formation of $NaAlH_4$ is incomplete, indicated by the presence of peaks related to NaH, Al and Na_3AlH_6. The Δp is more substantial with Ti as dopant than without dopant, in relation to the higher content of Na_3AlH_6 in the XRD pattern (peaks more intense). On the other hand, $NaAlH_4$ is the main phase after milling with chloride dopants, with NaCl and Al as products of their reduction as expected from the amplitude of the pressure drop. The efficiency of the dopants can be directly estimated from the milling curves by evaluating the time necessary to reach the maximum pressure decrease. The effectiveness of the dopants can be ordered as follows: $TiCl_3$>$CeCl_3$>$ScCl_3$>>Ti.

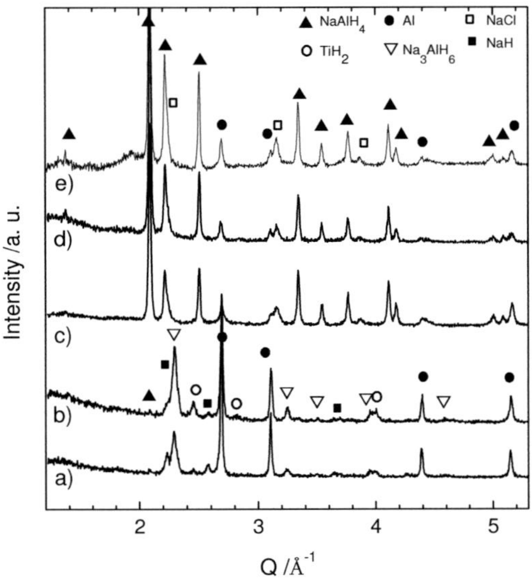

Figure 3.9: XRD patterns of the mixtures NaH + Al milled a) without dopant and with 4 mol% b) Ti, c) $TiCl_3$, d) $ScCl_3$ or e) $CeCl_3$

The results of the Rietveld refinement done on the XRD patterns obtained after milling with $TiCl_3$, $ScCl_3$ and $CeCl_3$ are given in **Table 3.2**. For all samples, the cell parameters are not significantly different from the bulk structure parameters given in the ICSD database (Inorganic Crystal Structure Database). This result indicates that the different phases are not deformed by the milling process in agreement with the low strain (<1%) determined for all phases. In addition, for the Al phase, the absence of cell deformation indicates that there is no substitution of metal (Ti, Sc or Ce) in the Al structure and formation of a solid solution. For the sample milled with $ScCl_3$, the fraction of $NaAlH_4$ determined from the XRD patterns is very close to the amount expected as for the sample milled with $TiCl_3$. By contrast, for the sample milled with $CeCl_3$, the amount of $NaAlH_4$ obtained is higher than expected that means that the decomposition of $CeCl_3$ is not complete during milling reducing the consumption of $NaAlH_4$ to form NaCl (equation (3.2)).

Table 3.2: Summary of the results obtained from Rietveld refinement of the XRD patterns for as-milled NaH + Al + 4 mol% dopant. The errors on the crystallite size and strains are about 10%.

Phases	*Dopant*	*Fraction (wt%) (expected[a])*	*a (Å) (reference[b])*	*c (Å) (reference[b])*	*Crystallite size (nm)*	*strain*
$NaAlH_4$	$TiCl_3$	79.9 (79.6)	5.0255 (5.0251)	11.3661 (11.3539)	53	0.0015
	$ScCl_3$	80.0 (79.7)	5.0253 (5.0251)	11.3608 (11.3539)	63	0.0017
	$CeCl_3$	78.6 (75.0)	5.0235 (5.0251)	11.3586 (11.3539)	80	0.0022
NaCl	$TiCl_3$	10.8 (11.7)	5.6326 (5.6440)	-	23	0.0040
	$ScCl_3$	13.5 (11.7)	5.6354 (5.6440)	-	26	0.0031
	$CeCl_3$	15.4 (11.1)	5.6360 (5.6440)	-	70	0.0058
Al	$TiCl_3$	9.2 (5.4)	4.0460 (4.0495)	-	26	0.0038
	$ScCl_3$	6.5 (5.4)	4.0587 (4.0495)	-	54	0.0025
	$CeCl_3$	6.0 (5.1)	4.0501 (4.0495)	-	98	0.0026

[a] Calculation considering the completion of the concomitants formation of $NaAlH_4$ and decomposition of XCl_3 (equation (3.2)). Balance for metallic elements Ti (3.2 wt%), Sc (3.0 wt%) and Ce (8.8 wt%).

[b] reference structures taken from the ICSD database with the PDF (Powder Diffraction File) numbers: 154907 ($NaAlH_4$), 041439 (NaCl) and 044321 (Al).

These results are already given by the observation of the pressure drops occurring during milling (**Table 3.1**), confirming that the study of the pressure variation recorded during milling gives valuable information and insights into the synthesis reaction. The fractions of Al and NaCl do not fit exactly the calculated values being an indication that the decomposition reactions of $ScCl_3$ and $CeCl_3$ differ from $TiCl_3$. The crystallite sizes give also important information for the comparison of the different dopants in the milling process. They are very small (<100 nm) as commonly obtained after the ball milling process, but some differences can be seen between the different dopants with the lowest size obtained for the sample milled with $TiCl_3$. This observation may be related to the higher efficiency of $TiCl_3$ during mechano-chemical synthesis of $NaAlH_4$ because it is established for hydrides that small grain sizes favour the H_2 diffusion and enhance the sorption kinetics [177].

3.1.4 Conclusions on the synthesis by reactive ball milling

A better understanding of the reactions occurring during milling can be reached by in-situ monitoring pressure and temperature during milling. It is first used to determine optimised milling conditions for a fast and efficient synthesis of doped-$NaAlH_4$. In addition, the method can validate the completion of the reaction and gives access to new information about the kinetics of multiple reaction steps. For example, the use of low and intermediate hydrogen pressures for the synthesis of $NaAlH_4$ induces, on the one hand, an increase of milling time compared to high pressure ball milling; but on the other hand, it allows for the first time depicting the sequence of reactions (1.6) and (1.7) already during synthesis. From all these results, it appears that a good synthesis performance (shorter milling time for reaction completeness) is obtained using 100 bar H_2 starting pressure, 37 balls and a rotational speed of 550 rpm. The addition of 4 mol% of dopant is a good compromise between kinetics and H_2 capacity. Moreover, the study of the addition of different dopants has disclosed $TiCl_3$ as the most efficient one for the synthesis of $NaAlH_4$. A slightly longer milling is necessary when using $ScCl_3$ or $CeCl_3$ but the synthesis completion can still be achieved in a reasonable time, whereas pure Ti would require a much longer milling time. The H_2 sorption properties of $NaAlH_4$ doped with different compounds is described in more detail in the next section.

3.2 Thermodynamic and kinetic properties of doped-$NaAlH_4$: comparison of dopants

In the previous section, the synthesis of doped-$NaAlH_4$ in one step by reactive ball milling is demonstrated. Different dopants were tested. For the synthesis process, it clearly appears that the chloride compounds are very efficient to promote the synthesis reaction. The aim of this section is to analyse in more detail the efficiency of these different compounds - Ti, $TiCl_3$, $ScCl_3$ and $CeCl_3$ - on the decomposition and re-absorption reactions.

3.2.1 Decomposition process

The decomposition of $NaAlH_4$ occurs in three steps described by reactions (1.6), (1.7) and (1.8). The sequence of decomposition is well established (see §1.2.1) and can be followed by in-situ synchrotron measurements (**Figure 3.10**) performed here for the sample milled with $CeCl_3$.

During heating to 90°C at 0.01 bar H_2, the patterns remain unchanged and are composed of $NaAlH_4$, NaCl and Al as obtained after milling. In addition, the presence of Na_3AlH_6 formed during the delay time between milling and measurement is noticed (see below). During further heating to 140°C, the $NaAlH_4$ phase completely disappears and the intensities of the peaks related to Na_3AlH_6 and Al increase. This change is quite abrupt confirming the fast kinetics of the first decomposition step of $NaAlH_4$ (equation (1.6)).

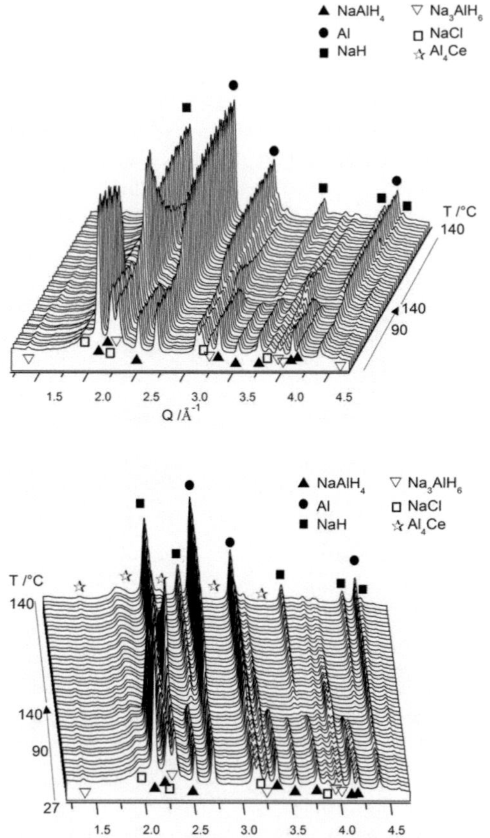

Figure 3.10: In-situ synchrotron measurements (two different views) during the first decomposition of as-milled $NaAlH_4$ + 4 mol% $CeCl_3$ upon heating to 140°C.

When the temperature reaches 140°C, new peaks appear that correspond to NaH; these peaks increase in the next patterns for the temperature kept at 140°C. At the same time, the peaks related to the Al phase continue to increase and the ones related to Na_3AlH_6 slowly decrease. These changes are related to the second step of decomposition (equation (1.7)) and are much slower than for the first step because the evolution of the peak increase/decrease can be followed in several patterns. This is related to the lower reaction rate of the second step already reported in literature (see §1.2.1). The formation of a new phase Al_4Ce is also noted during heating from the reaction between the decomposed dopant and Al. This last observation will be discussed in more detail in the following.

The decomposition of $NaAlH_4$ in different steps is also observed by differential scanning calorimetry (DSC), which can be used to compare doped and undoped samples (**Figure 3.11**). On the DSC trace measured for undoped $NaAlH_4$, four endothermic events occur (see §1.2.1). The first one (M) at ca. 180°C is related to the melting of $NaAlH_4$ and is followed by the decomposition of molten $NaAlH_4$ into Na_3AlH_6 (A), then of Na_3AlH_6 into NaH (B) and finally of NaH (C) [57]. On the other hand, in the DSC traces for doped-$NaAlH_4$ only three endothermic peaks arise during heating to 400°C related to the three steps of $NaAlH_4$ decomposition (A' for reaction (1.6), B' for reaction (1.7) and C' for reaction (1.8)).

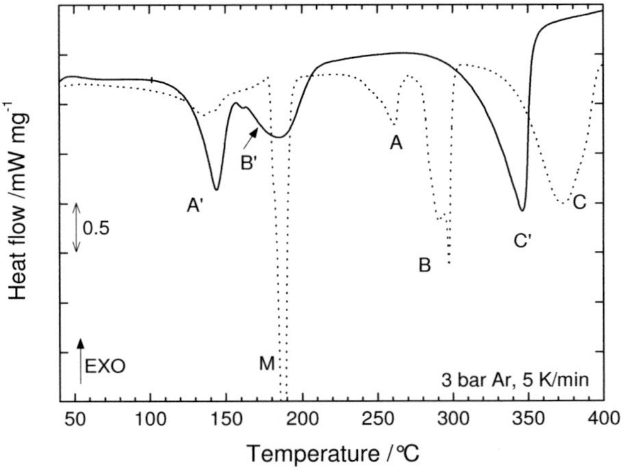

Figure 3.11: DSC traces at 5°C min^{-1} in Ar for undoped $NaAlH_4$ (dot line) and $NaAlH_4$ + 4 mol% $CeCl_3$ (solid line). The following events are observed: melting of $NaAlH_4$ (M), decomposition of $NaAlH_4$ (A, A'), decomposition of Na_3AlH_6 (B, B') and decomposition of NaH (C, C') according to reactions (1.6), (1.7) and (1.8), respectively.

3.2.2 Thermodynamic properties

In this section, the thermodynamic properties of $NaAlH_4$ are studied as well as the eventual effect of the addition of a dopant. If the effect of the dopants is purely catalytic, there should be no difference between the samples milled with different compounds for the enthalpy of reaction and the entropy that defines the equilibrium pressure-composition-temperature diagram. These values are obtained using equilibrium measurements (see §2.3.3) that are very time consuming and tedious. A method using isobaric DSC is proposed to approach the enthalpy and the entropy of the reaction using a non equilibrium method (for details see §2.3.1). The onset temperatures for H_2 desorption and absorption measured by isobaric DSC change as a function of the applied pressure. These values can be used to construct a van't Hoff plot from which boundary values of the enthalpy and the entropy can be derived. The experiments performed for $NaAlH_4$ milled with $TiCl_3$, $ScCl_3$ or $CeCl_3$ consist in cycling the samples by gradually increasing the H_2 pressure from 20 to 120 bar H_2. These experiments give then also information on the cycling ability of the different samples. The sequences of DSC traces obtained for the different dopants are given in **Figure 3.12**.

At first, the results obtained for the sample milled with $TiCl_3$ are described in detail (**Figure 3.12a**). The dotted trace under 20 bar was obtained after the complete sequence of cycles at different pressures to compare with the initial state. The two curves measured under 20 bar are very similar indicating the good reversibility of sodium alanate. On the different DSC traces, several phase transitions are observed during heating or cooling steps, depending on the hydrogen pressure. The peaks related to H_2 sorption shift to higher temperature with higher pressure, whereas the structural transitions always occur at the same temperature. The peaks are numbered from the trace under 120 bar but represent the same transitions for each pressure. As already shown above, endothermic peaks are observed during heating, whereas exothermic peaks are observed during cooling.

The different features observed during heating are identified as follows: 1) decomposition of $NaAlH_4$ (equation (1.6)), which can overlap with melting if the peak temperature is close to 180°C; 2) structural transformation of Na_3AlH_6 from α- to β-phase [48] and 3) decomposition of Na_3AlH_6 (equation (1.7)). Peaks 2 and 3 overlap on the DSC traces at low pressures. Considering reversible reactions, the peaks observed during cooling are related to: 4) formation of Na_3AlH_6, 5) structural transition of Na_3AlH_6 from β- to α-phase and 6) formation of $NaAlH_4$ which can also overlap with the solidification peak at approximately 180°C. Peaks 4 and 5 can also overlap at low pressures. Moreover, peak 6 appears clearly only for high pressures. For lower H_2 pressures, it is assumed that the peak of $NaAlH_4$ formation is very broad and thus has a weak intensity and that it is shadowed by the background of the curve. Additionally, a small exothermic "wave" often arises at a temperature lower than the one corresponding to peak 1. This feature can be related to hydrogen absorption during heating, when the hydride is not completely formed during cooling.

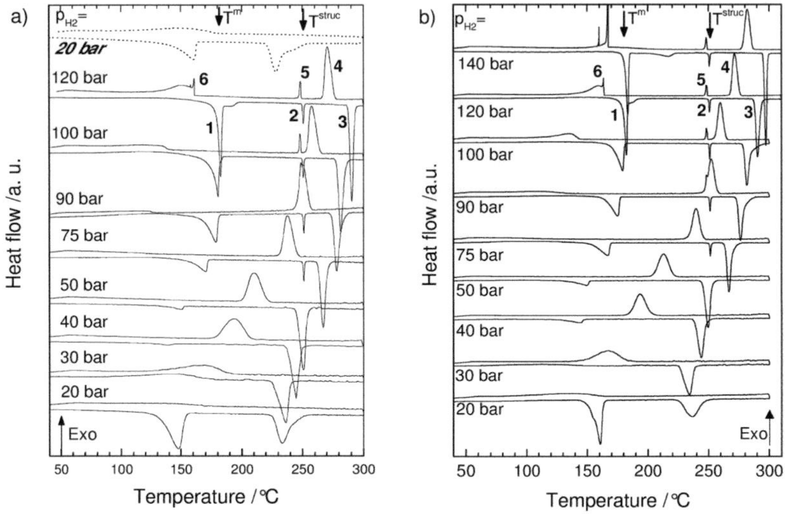

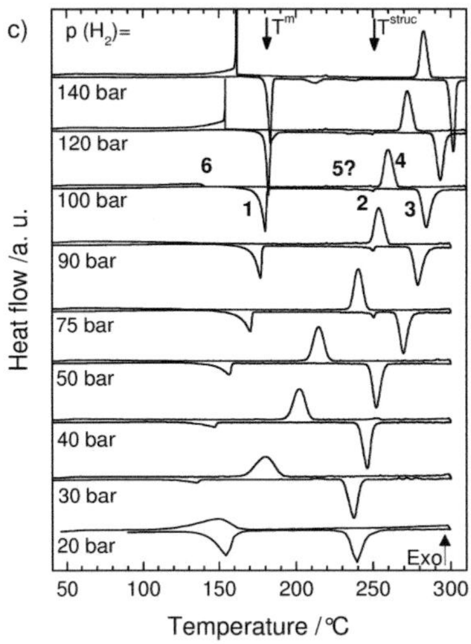

Figure 3.12: DSC traces (5°C min^{-1}) for doped-$NaAlH_4$ during cycling for increasing H_2 pressures. The dopants used are a) $TiCl_3$, b) $ScCl_3$ and c) $CeCl_3$. The downward arrows denoted T^m and T^{struc} give the temperatures of $NaAlH_4$ melting and the structural transformation of Na_3AlH_6, respectively.

An exception in the sequence of peaks described above is found for the trace measured under 30 bar H_2. In this case, the differences are the consequence of the incomplete re-absorption in the previous measurement under 20 bar. Thus, under 30 bar, the first event corresponds to an exothermic peak ascribed to hydride formation (like 4) and the second feature to an endothermic peak related to desorption (3). For higher H_2 pressure, re-absorption of hydrogen is observed during cooling and then two desorption peaks arise in the following cycle.

For the sample milled with $ScCl_3$, the sequences of the DSC traces for increasing H_2 pressure are similar to the results obtained with $TiCl_3$. The same events 1 to 6 occur for the same pressures. One difference is that, for the first trace at 20 bar H_2, it is possible to observe an exothermic peak during cooling that corresponds to the formation of Na_3AlH_6 from NaH. This result shows the better re-absorption properties of the sample milled with $ScCl_3$. In the next trace at 30 bar H_2, there is also a weak exothermic signal at the beginning of the heating stage that completes the previous reaction and then the decomposition peak corresponding to the decomposition of Na_3AlH_6 arises. The better absorption properties when milling with $ScCl_3$ are also illustrated at higher pressure for step 1 and 6 where the areas of the de-/absorption peaks (related to the amount of reacting material) is much larger than those for the DSC traces of the sample milled with $TiCl_3$. In addition for this sample, a DSC measurement was also performed at 140 bar H_2 where it is possible to see first the melting of $NaAlH_4$ at 180°C and the decomposition of liquid $NaAlH_4$ into Na_3AlH_6 at ca. 214°C.

For the sample milled with $CeCl_3$, similar observations can be made for the different traces. The re-absorption properties appear to be even better than for $ScCl_3$ as reflected by a higher absorption peak obtained during the cooling stage of the first cycle at 20 bar H_2 and the occurrence of the two decomposition steps during the following heating stage at 30 bar H_2. Similarly, the areas of events 1 and 6 are larger than those obtained when doping with $ScCl_3$ or $TiCl_3$. However, for $CeCl_3$, the very weak amplitude of peak 2 corresponding to the phase transformation of Na_3AlH_6 and the absence of the reverse reaction 5 are unexpected. The β-Na_3AlH_6 phase may not form during heating from the α-phase.

From the peak sequence obtained by DSC at different pressures, it is possible to construct the van't Hoff plot for absorption and desorption (**Figure 3.13**) with the inverse of the onset temperatures for each reaction as a function of the logarithm of the applied pressure. The values obtained follow linear trends but with different slopes for the different reaction steps. For the different dopants, the onset temperatures are similar for each reaction steps and the values in the van't Hoff plot are located on the same line. This is an indication that the dopants do not influence the thermodynamic properties of $NaAlH_4$. For the reaction step corresponding to the phase transition between Na_3AlH_6 and NaH (equation (1.7)), the energy range for enthalpy is between 33.6 ± 0.6 kJ mol-1 and 52.6 ± 0.8 kJ mol^{-1} for the absorption and desorption reactions, respectively. These values are in fact an average between the reaction steps from/to α- and β-Na_3AlH_6. From PCI measurements, Bogdanovic et al. [50] reported ΔH = 47 kJ mol^{-1} as an average between absorption and desorption values. Given that they used temperatures lower than 200°C, their results involved only α-Na_3AlH_6 but they can nevertheless be compared with the results obtained here. The enthalpy value of 47 kJ mol^{-1} appears between the results obtained here for absorption and desorption but the hysteresis is high because of the relative slow kinetics of reaction. This enthalpy range is acceptable knowing that, from calculations, the enthalpy is given with ± 10 kJ mol^{-1} error [55] similar to what is obtained from DSC experiments.

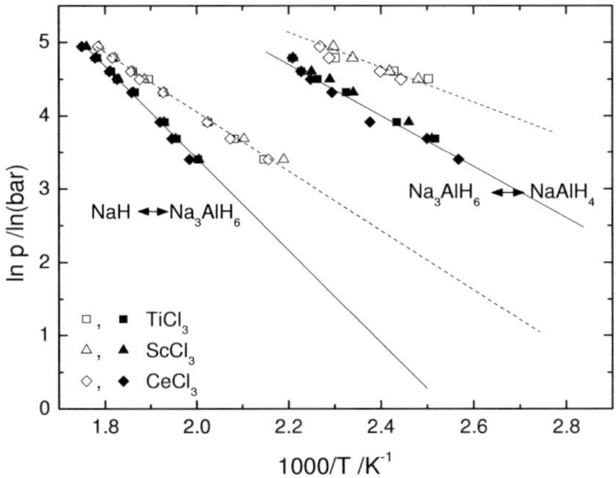

Figure 3.13: van't Hoff plots obtained from isobaric DSC for $NaAlH_4$ doped with 4 mol% $TiCl_3$, $ScCl_3$ or $CeCl_3$. Closed symbols: desorption; open symbols: absorption. The solid (desorption) and dotted (absorption) lines give the linear trend used for the determination of the reaction enthalpies.

For the reaction step between $NaAlH_4$ and Na_3AlH_6 (equation (1.6)), the energy range for the enthalpy is between 16.1 ± 1.7 kJ mol-1 and 28.0 ± 1.2 kJ mol^{-1} for the absorption and desorption reactions, respectively. Bogdanovic et al. reported for this step ΔH = 37 kJ mol^{-1} (average between absorption and desorption). The values obtained here are slightly smaller from this enthalpy but are not far regarding the fact that this reaction is much more difficult to study and even close to values given by different authors (see **Table 1.3**). High pressures are necessary to obtain this reaction, $NaAlH_4$ is relatively instable and kinetics are slow. Moreover, the corresponding peaks often overlap with melting or solidification peaks like for the trace under 120 bar H_2. It is therefore difficult to determine the onset temperature. Overall, the results obtained by DSC measurements are in the order of those reported for $NaAlH_4$ and demonstrate that careful isobaric DSC measurements can be used as a valuable tool to approximate the enthalpy values.

Conventional PCI experiments were also performed to compare the samples prepared with different dopants and also to evaluate the validity of the isobaric DSC measurements. The PCI curves performed at 150°C for absorption and desorption (**Figure 3.14**) show two plateaus, at low and at high pressure, corresponding to reactions (1.6) and (1.7), and an important hysteresis between the absorption and desorption curves. This hysteresis is typical of kinetic limitations, here caused by the relatively low reaction rates that do not allow to reach equilibrium with the time allotted for each pressure step (see §2.3.3). Nevertheless, the equilibrium pressures are in agreement with PCI previously reported (**Figure 1.5**) and are similar for all dopants for absorption and for desorption in both reaction steps. This result confirms that the dopants used affect only the kinetics of the reaction and do not modify the thermodynamic properties of $NaAlH_4$.

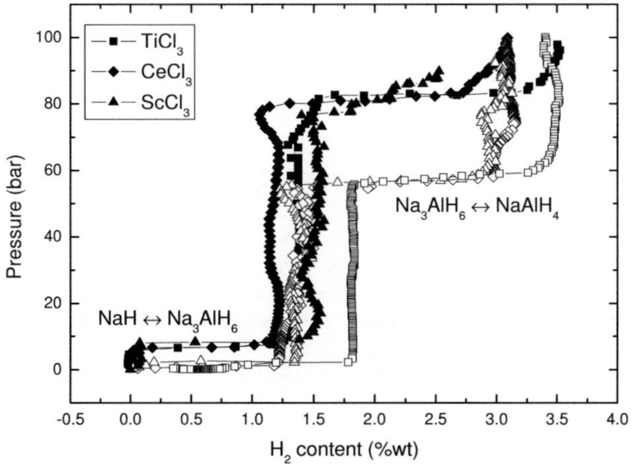

Figure 3.14: PCI curves measured at 150°C for $NaAlH_4$ doped with $TiCl_3$, $ScCl_3$ or $CeCl_3$. Closed symbols for absorption, open symbols for desorption.

The values of equilibrium pressure can be transferred to van't Hoff plots (**Figure 3.15**) and compared with the results obtained by DSC represented by straight lines. The values obtained by PCI measurements are located close to or between the lines plotted from DSC confirming that these last ones enclose the real enthalpy values. For the reaction from Na_3AlH_6 to $NaAlH_4$, an enthalpy of 27.6 ± 0.8 kJ mol^{-1} is calculated that is also in the range previously reported in the literature (**Table 1.3** and references therein). For the reaction between NaH and Na_3AlH_6, the low precision of the pressure sensor at low pressures (technical limitations) induces large errors in the determination of the equilibrium pressure and the values obtained by fitting the van't Hoff plot are not relevant.

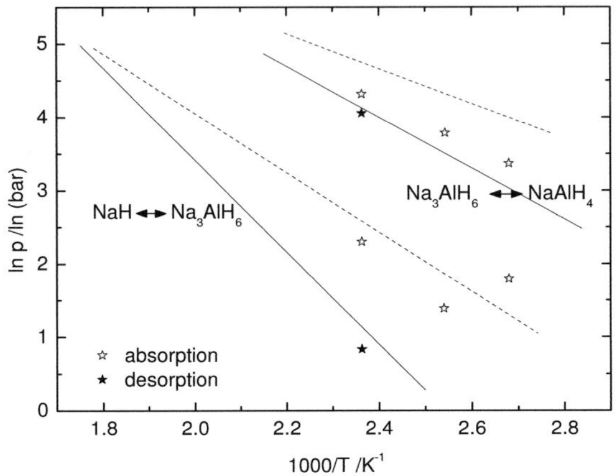

Figure 3.15: van't Hoff plots obtained from PCI curves measured at 100, 120 and 150°C for $NaAlH_4$ doped with 4 mol% $TiCl_3$ in absorption (open symbols) and in desorption (closed symbols) and comparison with the linear trends obtained from isobaric DSC (Figure 3.13) in desorption (solid lines) and in absorption (dotted lines).

3.2.3 Kinetic properties

3.2.3.1. Decomposition

The first step in the analysis of the kinetic properties is the study of the decomposition of the as-milled samples. The comparison for different dopants can be done by DSC measurements of the decomposition in Ar atmosphere. As shown in §3.2.1, the decomposition of $NaAlH_4$ starts at a much lower temperature using chloride dopants than for pure $NaAlH_4$, before melting. Differences between the DSC traces measured for $NaAlH_4$ milled with different dopants are observed (**Figure 3.16**). The onset temperature of the first peak (A) is lower when using $TiCl_3$ (110°C) than when using $ScCl_3$ or $CeCl_3$ (125°C), in agreement with the higher efficiency of $TiCl_3$ observed during the milling. On the other hand, the second decomposition step (B) starts earlier for the sample milled with $CeCl_3$ than with $TiCl_3$ because of a shorter delay between the two reaction steps. This result indicates a higher efficiency of $CeCl_3$ for the decomposition of Na_3AlH_6. $ScCl_3$ is the less efficient dopant for the decomposition of $NaAlH_4$ and Na_3AlH_6. For the sample milled with metallic Ti, only peak B is observed because only Na_3AlH_6 was obtained by ball milling. In addition, the onset temperature is higher for peak B than for the chloride dopants confirming the lower efficiency of pure Ti as dopant.

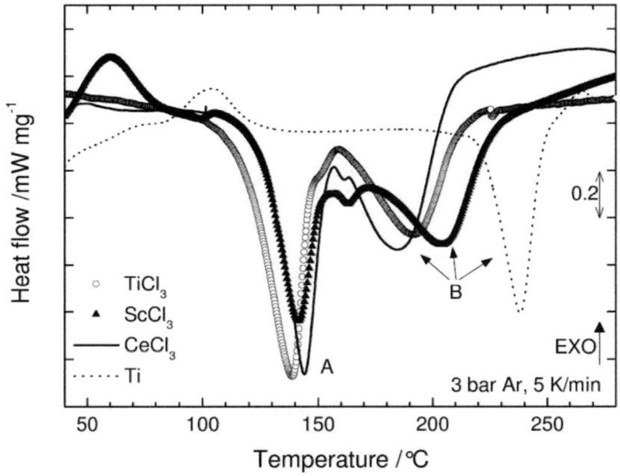

Figure 3.16: DSC traces for as-milled $NaAlH_4$ with different dopants measured in Ar. Event A corresponds to the decomposition of $NaAlH_4$ into Na_3AlH_6 and event B of Na_3AlH_6 into NaH.

The H_2 release of the as-milled samples was also analysed by thermogravimetric (TG) measurements. The results are in agreement with the DSC results: the mass losses are related to the H_2 release according to the three endothermic reactions (1.6) to (1.8) seen in the DSC traces for $TiCl_3$ (**Figure 3.17a)**. The first two decomposition steps follow each other without delay and can be distinguished by the change of mass loss rate in the TG curve, indicating a lower reaction rate for the second reaction step (see §1.2.1). The comparison of the TG curves for the doped and for the undoped samples (**Figure 3.17b**) confirms that the decomposition of $NaAlH_4$ occurs at a much lower temperature when using chloride dopants. For the undoped $NaAlH_4$, the two first steps appear to occur in a single one with mass loss starting at ca. 240°C. For the sample doped with pure Ti, only the mass losses related to the decomposition of Na_3AlH_6 and then of NaH are observed because only Na_3AlH_6 was formed during milling (**Figure 3.9**). Among the chloride dopants, the lowest decomposition temperature is obtained for $TiCl_3$ (**Annexe D**).

For all samples, the H_2 capacities measured – step (1.6) and (1.7) - are smaller than calculated from the total amount of $NaAlH_4$ present in the samples after milling (see **Annexe D**). A first explanation is the low purity of the starting products $NaAlH_4$ (90%) or NaH (95%) that decreases the fraction of active compound in the powder. In addition, for the samples doped with chloride compounds, their very high efficiency enables the decomposition of $NaAlH_4$ already during storage in the glove box (T_{dec} at 30°C at 1 bar H_2, see §1.2.1) so that the samples are no more fully hydrogenated when starting the thermogravimetry measurements some days after milling (longer delay for sample doped with $CeCl_3$).

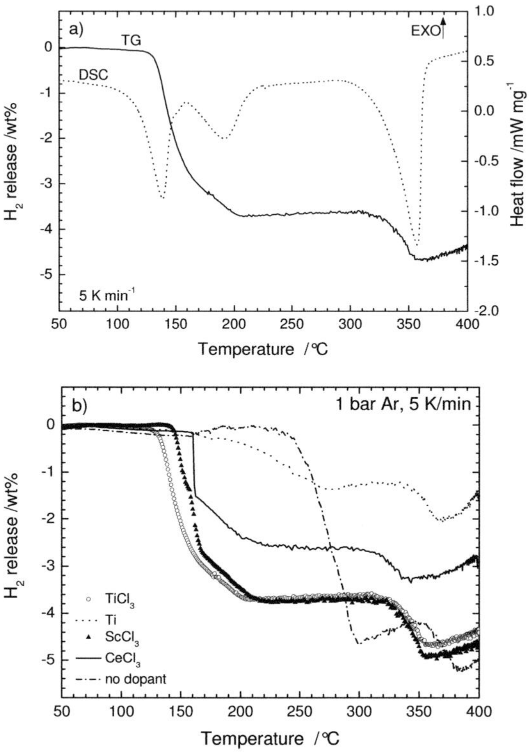

Figure 3.17: a) Comparison of DSC traces (3 bar Ar) and TG curves (1 bar Ar) for $NaAlH_4$ milled with 4 mol% $TiCl_3$ and b) TG curves (1 bar Ar) for undoped and doped $NaAlH_4$ (4 mol% dopant, 5°C min^{-1})

The decomposition of $NaAlH_4$ at room temperature is demonstrated by the evolution of the XRD patterns of the different samples after long time storage in the glove box (**Figure 3.18**). The relative efficiency of the dopants can be found here again because the sample doped with $TiCl_3$ shows the appearance of Na_3AlH_6 after less days compared to $NaAlH_4$ doped with $ScCl_3$ or $CeCl_3$. This decomposition is fully reversible.

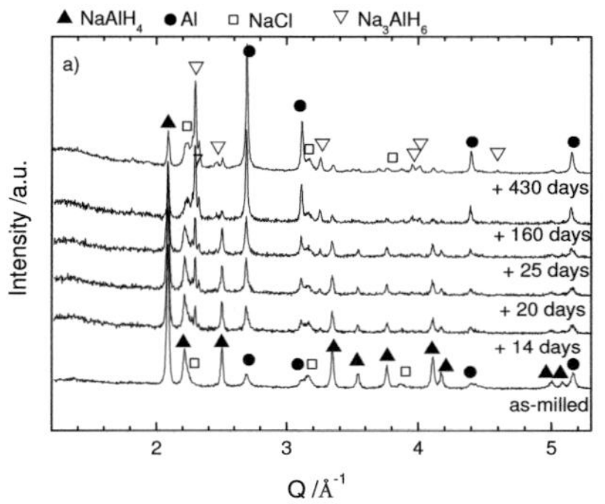

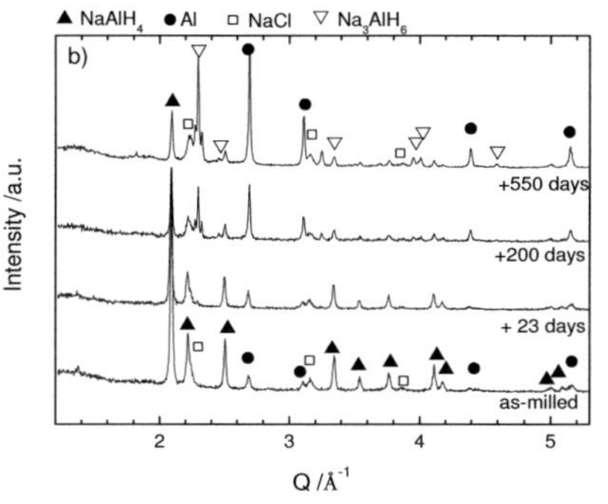

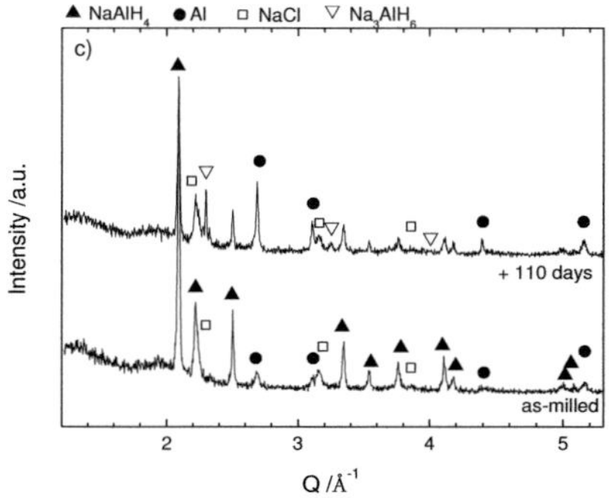

Figure 3.18: XRD patterns measured after different storage time in the glove box for as-milled $NaAlH_4$ with 4 mol% a) $TiCl_3$, b) $ScCl_3$ and c) $CeCl_3$

3.2.3.2. Re-absorption and cycling

From the analysis of the synthesis of $NaAlH_4$ and the decomposition of the as-milled samples, it appears that $TiCl_3$ is the more efficient dopant used. However, the advantages of $ScCl_3$ or $CeCl_3$ have been shown for the re-absorption of hydrogen and upon cycling [77] (**Figure 3.12**). This effect can be verified by DSC under H_2 pressure (**Figure 3.19**). A first cycle was performed at 3 bar H_2 to decompose $NaAlH_4$ during heating to 300°C considering only the two first steps of decomposition (equation (1.6) and (1.7)) useful for storage. During heating, two endothermic peaks A and B arise for all samples and correspond to the two steps of decomposition. They show higher onset temperatures (125-130°C) when measured in 3 bar H_2 than when measured under Ar (**Figure 3.16**), in agreement with the van't Hoff equation (equation (2.8)). The DSC traces during heating are similar for all dopants because the kinetics do not govern the desorption process at 3 bar H_2 but rather the thermodynamics. During cooling, no peaks can be seen at 3 bar H_2 because the temperature necessary for re-absorption is too low for the reaction to proceed during the DSC measurements (slow kinetics and low precision, **Figure 1.5**).

In the second cycle, the H_2 pressure is set at 50 bar H_2 where the re-absorption should be obtained below the equilibrium temperatures at 50 bar H_2, which are at ca. 110 and 210°C for equations (1.6) and (1.7), respectively. In this cycle, different results are obtained for $NaAlH_4$ doped with $TiCl_3$, $ScCl_3$ or $CeCl_3$. For the sample doped with $TiCl_3$, the first event is an exothermic peak B' occurring during heating, followed by an endothermic one B. The exothermic peak B' corresponds to the re-absorption of hydrogen from NaH (obtained after the first cycle) to Na_3AlH_6. The temperature is too high after B' to allow the formation of $NaAlH_4$ from Na_3AlH_6 (equilibrium at 110°C) but Na_3AlH_6 decomposes during further heating (B) at higher temperature than at 3 bar H_2 following the van't Hoff equation.

For the sample doped with $ScCl_3$, the exothermic re-absorption B' occurs at lower temperature than for $TiCl_3$ but still too high for the second absorption step, whereas two steps of re-absorption (two maxima in the exothermic peaks B' and A') are obtained at the lowest temperature for the samples doped with $CeCl_3$ (**Annexe D**). During further heating, the two decomposition steps of the newly formed $NaAlH_4$ take place corresponding to the two endothermic events A and B observed at ca. 140 and 245°C. These results indicate that $ScCl_3$ and $CeCl_3$ are better dopants for the absorption step than $TiCl_3$.

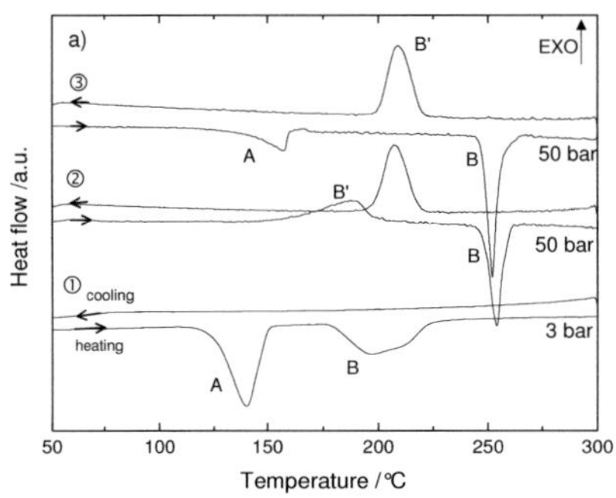

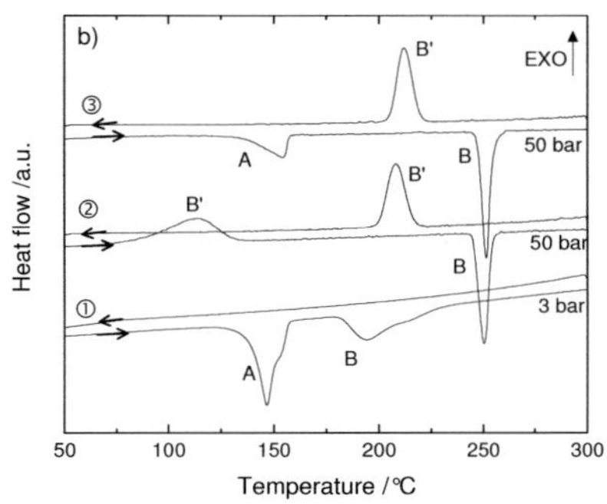

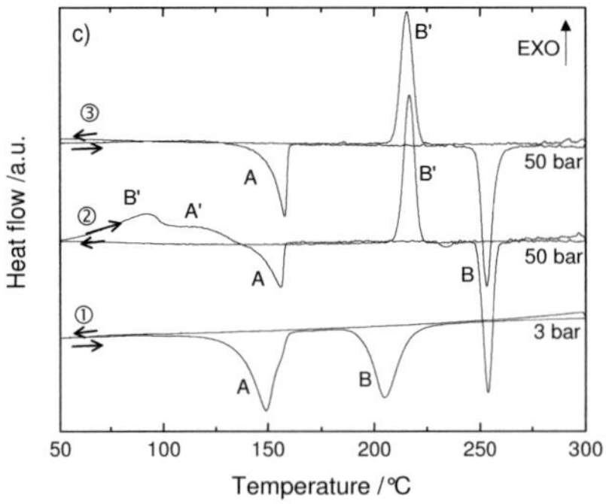

Figure 3.19: DSC traces for cycle ① at 3 bar H_2 and cycles ② and ③ at 50 bar H_2 for the as-milled $NaAlH_4$ with 4 mol% a) $TiCl_3$, b) $ScCl_3$ and c) $CeCl_3$

A : $NaAlH_4 \rightarrow Na_3AlH_6$

A': $Na_3AlH_6 \rightarrow NaAlH_4$

B : $Na_3AlH_6 \rightarrow NaH$

B': $NaH \rightarrow Na_3AlH_6$

During the cooling stage of the second cycle and the third cycle, still at 50 bar H_2, the DSC traces are similar again for all samples. The cooling part of the second cycle is composed of a large re-absorption peak B' corresponding to the formation of Na_3AlH_6 from NaH. It is not possible to distinguish the second absorption step from Na_3AlH_6 to $NaAlH_4$ at lower temperature likely because the kinetics are too slow so that the corresponding exothermic peak is hidden in the background. However, during the third cycle, the two endothermic peaks A and B are again observed that are related to decomposition, confirming the complete reformation of $NaAlH_4$ during previous cooling.

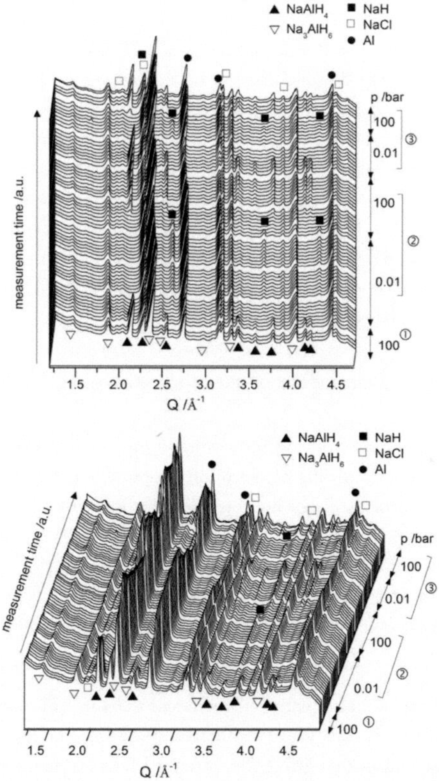

Figure 3.20: In-situ synchrotron measurements for sample $NaAlH_4$ milled with 4 mol% $CeCl_3$ (two different views) during cycling at 140°C by changing the applied H_2 pressure (cycle ① to ③). Continuation of the experiment shown in Figure 3.10 (cycle ①). Al_4Ce phase is not shown in the figure to avoid crowding the graphs but is present in all patterns.

The high efficiency of $CeCl_3$ for re-absorption is also shown by synchrotron measurements performed upon cycling (**Figure 3.20**), especially for the first absorption step from NaH to Na_3AlH_6. The experiment shown in **Figure 3.20** starts after the first decomposition (**Figure 3.10**) followed by 2.5 h at 100 bar H_2. Despite this rather long absorption time, there is still a significant amount of Na_3AlH_6 in the sample at the beginning of the experiment. This is related to the high equilibrium pressure (ca. 50 bar, see **Figure 1.5**) of the absorption reaction from Na_3AlH_6 to $NaAlH_4$ that implies a low driving force for the reaction at 140°C and therefore slow kinetics. During the experiment, the cycling ability of doped-$NaAlH_4$ is evaluated by changing the pressure between 0.01 and 100 bar H_2 at a constant temperature of 140°C.

After the first change to 0.01 bar, the decompositions of $NaAlH_4$ and then Na_3AlH_6 into NaH and Al are observed as described above for the first desorption (**Figure 3.10**). There is always some Na_3AlH_6 left in the sample since the measurements are done for a short period of time for each step (total time for the experiment: 50 min) and, as already observed, the decomposition of Na_3AlH_6 is rather slow. For the second re-absorption, the disappearance of NaH, decrease of Al and growth of Na_3AlH_6 and $NaAlH_4$ is very fast since it can be seen in the same scan where the high pressure (100 bar H_2) is applied (exposure time 30s). The second absorption step from Na_3AlH_6 to $NaAlH_4$ is slower as reflected by the more continuous increase/decrease of the different phases. The Al_4Ce phase (not shown in **Figure 3.20**) formed during the first decomposition remains unchanged during further cycling.

3.2.4 Summary of the dopant effect on thermodynamic and kinetic properties

From all the results obtained by studying the different dopants, the following conclusions about the effect of these compounds on the H_2 sorption properties of $NaAlH_4$ are drawn. The most efficient dopants found during synthesis are the chloride compounds $TiCl_3$, $ScCl_3$ and $CeCl_3$. The analysis of the thermodynamic properties shows that the chloride dopants have no influence on the thermodynamics and that their effect is purely kinetic. Using pure Ti as a dopant leads to poor kinetics indicating that not only the metal is important for the catalytic mechanism but also the Cl atoms. Nevertheless, the relative efficiency of the dopants is different for the desorption or the absorption processes.

The kinetics of desorption are better with $TiCl_3$ whereas the absorption rate is faster using $CeCl_3$. The aim of the next section is therefore to understand the catalytic mechanisms for desorption and absorption that can explain the differences observed between the dopants. It will also clarify the role of the different elements (Ti, Sc and Ce) and of the chlorine atom on the catalytic mechanisms and attempt to define what the real catalytic species is.

3.3 Catalytic mechanism

It has been described in §1.2.1 that the catalytic mechanism in $NaAlH_4$ is not exactly known. Nevertheless, it has been demonstrated that Ti acts as a catalyst for the dissociation of H_2 molecules at the surface and then facilitates H absorption in $NaAlH_4$ [114]. However, this step is not the limiting step of the reaction that appears to be more related to a mass transfer process in $NaAlH_4$. This statement is well seen by comparing the effect of Ti and $TiCl_3$ on $NaAlH_4$ kinetic properties. Ti allows improving the H_2 sorption reaction rates compared to undoped $NaAlH_4$ but it is far less efficient than $TiCl_3$. From these results, the decomposition of the dopant during milling and the formation of NaCl are assumed to play an important role in the reaction rate enhancement. The comparison of $NaAlH_4$ prepared with different chloride dopants showing a different efficiency can help to understand the catalytic mechanism. The Sc and Ce compounds are known species for catalysis (e.g. CeO_2 [178]) and they may split the H_2 molecule as efficiently as Ti. The difference observed between the different dopants may therefore be related to the mass transfer step. In addition, it is likely that the catalytic mechanisms are different for absorption and desorption since the best dopant for each step is different.

3.3.1 Catalytic mechanism for the H_2 desorption

3.3.1.1. Analysis of decomposition products

For $NaAlH_4$ doped with $CeCl_3$, synchrotron measurements show the formation of Al_4Ce during the first decomposition step. Metallic Ce should be formed during milling by the decomposition of $CeCl_3$ indicated by the presence of NaCl. Metallic Ce then alloys with Al during heating. XRD measurements were performed on the decomposed samples (after TG measurements, **Figure 3.17b**) doped with other compounds to verify the formation of similar phases. The XRD patterns (**Figure 3.21**) show that all samples have completely decomposed since no trace of hydride phases: $NaAlH_4$, Na_3AlH_6 or NaH are seen. In addition, new phases appear during decomposition, involving the dopant: TiH_2, Al_3Ti, Al_xSc_{1-x} and Al_4Ce for the samples doped with Ti, $TiCl_3$, $ScCl_3$ and $CeCl_3$, respectively.

The formation of Al_3Ti has been already reported for $TiCl_3$ doped-$NaAlH_4$, but, directly added to $NaAlH_4$, it has little influence on the catalytic mechanism [73, 98, 109]. On the contrary, the formation of TiH_2 has been reported to favour especially the re-absorption of hydrogen in the sample [77, 179]. The formation of Al_xSc_{1-x} alloys is mainly seen in the XRD pattern from the asymmetry of the Al peaks towards lower Q values. The substitution of Sc in Al leads to larger cell parameters, in contrast to the other transition metals [180]. The formation of Sc-Al phases has also been observed by NMR [94]. The asymmetry is more pronounced after decomposition than after milling (although the Al cell parameter determined by Rietveld refinement is slightly higher than for pure Al in the as-milled sample, see **Table 3.2**) indicating that this phase is formed during heating as observed for Al_3Ti or Al_4Ce. The formation of Al_3Sc in Sc-doped

$NaAlH_4$ has been found to decrease the hydrogen desorption energy [92], compared to undoped $NaAlH_4$. A similar effect is possible for such Al_xSc_{1-x} phases. The formation of Al_4Ce has already been shown in the previous section. It has also been observed by other authors and this compound shows a very high efficiency for the H_2 absorption [133, 135]. Nevertheless, Al_4Ce contains more metallic Al than available after the decomposition of $CeCl_3$ by the reaction with $NaAlH_4$ (4 atoms necessary for 3 available, equation (3.2)) meaning that there is consumption of Al from the decomposition of $NaAlH_4$ that can limit the capacity obtained during re-absorption.

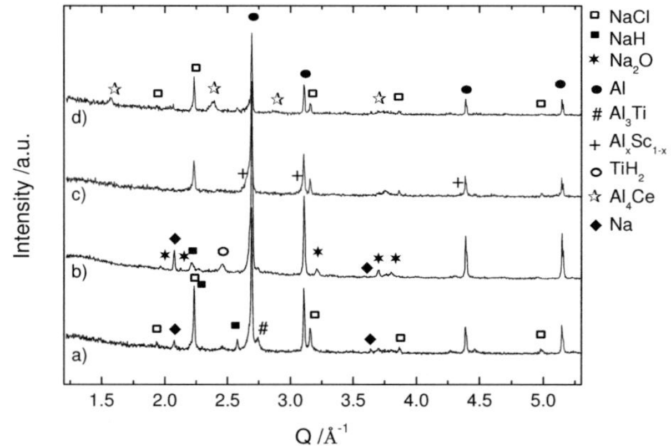

Figure 3.21: XRD patterns of the decomposed samples (after TG measurements, heating to 400°C in 1 bar Ar) doped with a) $TiCl_3$, b) Ti, c) $ScCl_3$ and d) $CeCl_3$

The possible influence on the catalytic mechanism of the different Al-M (M=Ti, Sc, Ce) phases formed when doping with MCl_3 will be discussed in more detail below. On the XRD patterns obtained after long time storage (**Figure 3.18**), the presence of phases related to the dopant like Al_4Ce or Al_3Ti cannot be detected, whereas they are obtained after the first decomposition step. This result indicates that these compounds are formed during heating and are not directly related to the decomposition of the dopant occurring during milling.

3.3.1.2. Morphology of decomposed samples

The morphology of as-milled and decomposed samples prepared with $TiCl_3$ and $CeCl_3$ was observed by SEM (**Figure 3.22** and **Figure 3.23**). A first observation is that the morphologies of the as-milled samples are really similar for both samples with large aggregates composed of smaller particles of some micrometers diameter.

Afterwards, the as-milled samples were heated ex-situ to 300°C in Ar atmosphere to fully decompose them. A change of the powder morphology can be clearly seen after heating. The sample milled with $TiCl_3$ (**Figure 3.22c-d**) shows much larger aggregates (several hundreds of micrometers) than before heating and they are more compacted. The size of the particles (1-10 µm) composing the aggregates is not dramatically changed but these particles show a very rough surface with flat islands (<1µm) on it. EDX mapping (not shown) of the decomposed sample indicates that no significant segregation of the different elements in different particles occurred during heating. Considering the resolution of EDX (ca. 1µm^3), a segregation of the elements within one particle cannot be excluded.

Figure 3.22: SEM pictures of sample $NaAlH_4$ doped with 4 mol% $TiCl_3$ at different magnifications: a+b: as-milled sample and c+d: after heating to 300°C in Ar atmosphere

For the sample milled with $CeCl_3$ (**Figure 3.23c**), much smaller (ca. 10 µm) and less compacted aggregates are observed. They are composed of particles with a similar size as after milling. However, using higher magnification (**Figure 3.23d**), the surface appears much rougher than for $TiCl_3$ with the presence of small round nodules (100-200 nm) that are "sintered" on the particle itself. The morphology observed by SEM can be related to the results of the BET measurements made on the same samples as-milled and after heating in-situ to 300°C. The surface areas for both as-milled samples are really similar with 8.3 $m^2\ g^{-1}$ for the sample milled with $TiCl_3$ and 9.6

$m^2\ g^{-1}$ for the sample with $CeCl_3$ (errors on surface area $\pm$ 1 $m^2\ g^{-1}$). Assuming that the powders are composed of spherical particles, these areas correspond to particles of 500-600 nm diameter, consistent with the average particle size observed by SEM (**Annexe D**).

After decomposition, different values are obtained for the surface areas: 6.6 $m^2\ g^{-1}$ for the sample milled with $TiCl_3$ and 2.8 $m^2\ g^{-1}$ for the sample milled with $CeCl_3$. Considering the error intrinsic to the BET measurement (hysteresis between desorption and adsorption of N_2), the surface area of the sample milled with $TiCl_3$ is considered largely unchanged after decomposition (average spherical diameter 700 nm). However, there is a significant surface area decrease for the sample milled with $CeCl_3$ (3 times lower) that corresponds to larger particles (average spherical diameter 1.7 μm), in agreement with the sintered particles observed by SEM.

Figure 3.23: SEM pictures for $NaAlH_4$ with 4 mol% $CeCl_3$ at different magnifications. a+b: as-milled sample and c+d: after heating to 300°C in Ar atmosphere.

The morphological differences have probably a direct influence on the kinetic properties of $NaAlH_4$. It is well known that hydrogen diffusion is faster in nano-structured hydrides (e.g. MgH_2) because of the smaller diffusion paths [181]. In addition, it has also been shown that confining $NaAlH_4$ in a nano-sized structure, i.e. by loading $NaAlH_4$ in a nano-structured framework, largely decreases the hydrogen desorption temperature [182]. Therefore, the reduction of the diffusion paths may explain the faster kinetics of H_2 desorption observed for $TiCl_3$. This reduction is shown by the smaller particle sizes maintained during heating shown by SEM and BET and by smaller crystallised domain sizes determined by Rietveld refinement for the sample milled with $TiCl_3$ (**Table 3.2**).

3.3.1.3. XPS analysis

Another difference found between the different dopants is that $CeCl_3$ is not completely decomposed during milling (**Table 3.1**). The previous discussion about the difference between the effect of Ti and $TiCl_3$ leads to the idea that the reduction of the dopant to the metal element could play an important role in the catalytic mechanism. X-ray photo-electron spectroscopy (XPS) was carried out to investigate the chemical state of the dopant in the as-milled samples. **Figure 3.24** shows the XPS spectra obtained for as-milled samples (with $TiCl_3$, $ScCl_3$ and $CeCl_3$ as dopant) before and after sputtering the surface.

In the Ti 2p spectrum obtained for the sample doped with $TiCl_3$ (**Figure 3.24a**), two large peaks are observed before sputtering for the Ti $2p_{3/2}$ and Ti $2p_{1/2}$ levels at 455.6 eV and 459.5 eV, ascribed to the presence of Ti^{II} species [183]. Here, the presence of near-surface TiO is confirmed by the O 1s signal found during XPS analysis (not shown). After sputtering the surface, the Ti $2p_{3/2}$ peak is shifted to the metallic state energy at 453.8 eV indicating that only metallic Ti is present beneath the surface. This observation is in agreement with results previously reported by Léon et al. [95]. It also confirms that $TiCl_3$ was completely reduced to metal Ti as assumed from the total pressure drop during milling (**Table 3.1**). The XPS analysis does not allow to discriminate whether Ti is present as pure metal in the sample or is forming an alloy with Al as reported by Bellosta von Colbe et al. [175].

On the other hand, for $NaAlH_4$ doped with $ScCl_3$, the surface is still partly composed of $ScCl_3$ even after sputtering. In the spectra of the Sc $2p_{3/2}$ and Sc $2p_{1/2}$ levels (**Figure 3.24b**), three peaks are observed: at 401.6 eV, which is ascribed to Sc_2O_3 (shoulder) [184], at 403.5 eV and 407.6 eV to $ScCl_3$ [185]. After sputtering, a new peak appears at 398.5 eV, corresponding to metallic Sc, together with the remaining peaks of $ScCl_3$ and Sc_2O_3 (weak). The pressure variations observed during milling and the XRD analysis indicate an almost complete reduction of $ScCl_3$ (**Table 3.1**). From the XPS analysis, it appears that at least the surface of the particles still contains $ScCl_3$. Bowman et al. [94] have also observed that $ScCl_3$ is not completely decomposed after milling since they still detect its signal by NMR. Considering the overlap between the Sc $2p_{3/2}$ peak of $ScCl_3$ and the Sc $2p_{1/2}$ peak of Sc metal, it is not possible to determine the relative amount of $ScCl_3$ and Sc. Nevertheless, the presence of $ScCl_3$ is obvious and evidences that part of the dopant (probably small) is still present in the sample after milling.

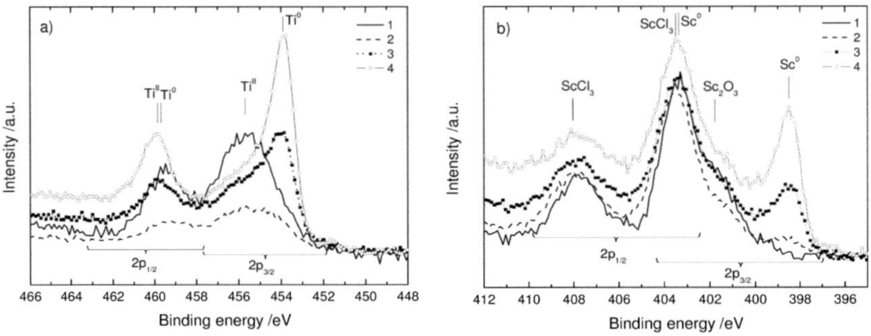

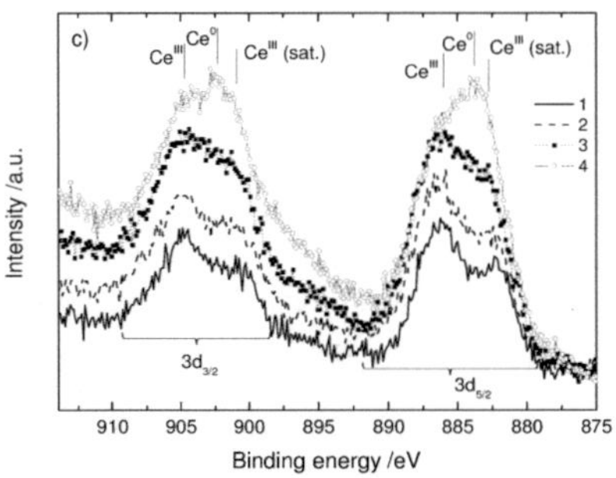

Figure 3.24: XPS spectra for a) Ti 2p, b) Sc 2p and c) Ce 3d of the surface of as-milled doped-$NaAlH_4$ (curve 1, solid line) and after sputtering for 2 min (2, dash line), 7 min (3, black squares) and 17 min (4, open circles).

The surface of the sample $NaAlH_4$ doped with $CeCl_3$ is only composed of Ce^{III} species (**Figure 3.24c**). The XPS spectrum of the Ce $3d_{5/2}$ level shows the presence of two peaks at 882.2 and 886.2 eV that are related to the Ce^{III} state (Ce_2O_3 or $CeCl_3$) and a shake-down (satellite) of this state [186], respectively. The main peak of Ce^{III} is still observed after sputtering together with an increasing peak at 883.5 eV related to the metallic state. The presence of Ce_2O_3 cannot be excluded; however, most of the oxide layer was removed by sputtering in the case of $ScCl_3$ or $TiCl_3$ as dopant. Thus, the presence of remaining $CeCl_3$ and metallic Ce is consistent with the pressure variations during milling that show that $CeCl_3$ has been only partly decomposed during milling.

XPS analyses demonstrate that $TiCl_3$, $ScCl_3$ and $CeCl_3$ are reduced to metal during milling. However, $ScCl_3$ and $CeCl_3$ but no $TiCl_3$ are present at the surface. The changes of the surface composition before and after sputtering (**Annexe D**) show that Ti, Sc and Al are found in larger amount after sputtering indicating that they are more present beneath the surface. The same observation can be made for Al and indicates a possible relation between the metals and Al. By contrast, the content of Cl increases with the sputtering time whereas the quantity of Na decrease meaning that no direct correlation is present between the two elements despite the formation of NaCl during milling. This result indicates that the surface is likely composed of sodium oxides.

The evolution of the chemical state of the dopant during desorption was studied by the analysis of milled samples that have been heated to 300°C under high vacuum in the preparation chamber of the XPS apparatus. A mass spectrometer attached to the chamber allowed checking the release of H_2 upon heating to ensure the completion of $NaAlH_4$ decomposition. It was also possible to see the release of O_2 and H_2O likely originating from surface contamination. The XPS spectra obtained for Ti 2p, Sc 2p and Ce 3d after heating are given in **Figure 3.25**.

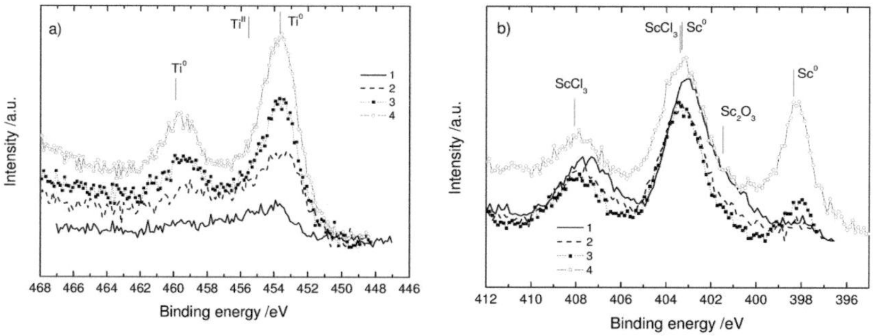

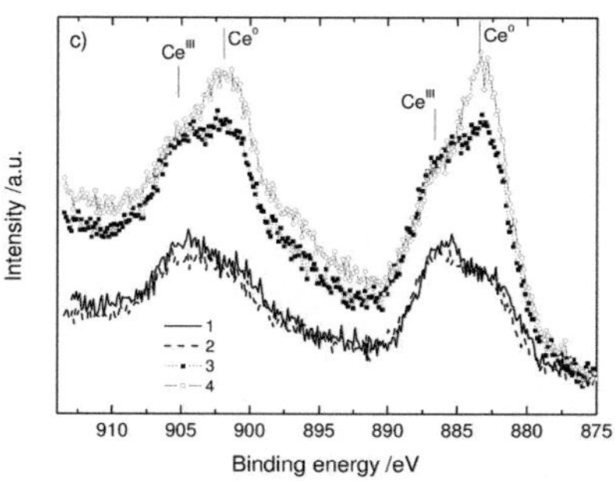

Figure 3.25: XPS spectra for a) Ti 2p, b) Sc 2p and c) Ce 3d for the surface of doped-$NaAlH_4$ (curve 1, solid line) after heating in-situ to 300°C and after sputtering for 2 min (2, dash line), 7 min (3, black squares) and 17 min (4, open circles).

The Ti 2p spectrum of the raw surface shows that Ti is only present in the metallic state after heating (**Figure 3.25a**). This confirms that TiO detected for the as-milled sample comes from surface contamination that was removed during heating in vacuum. Further sputtering of the surface only indicates an increase of the Ti content beneath the surface as already noticed in the as-milled sample. The spectrum of Sc 2p remains unchanged after heating with the presence of $ScCl_3$ and Sc_2O_3 (**Figure 3.25b**). This oxide is not removed by heating (more stable than TiO, ΔHf^0= -519 and -1908 kJ mol^{-1}, for TiO and Sc_2O_3, respectively [187]) but disappears during sputtering. In addition, during sputtering, the intensity of the peak related to metallic Sc increases significantly. Metallic Sc may alloy to Al in the form of Al_xSc_{1-x} as observed by XRD after the first decomposition (**Figure 3.21**).

For Ce 3d, there are also few differences before and after heating (oxide and chloride compounds are very stable) and in the same way the peaks of metallic Ce appear and grow with the sputtering time (**Figure 3.25c**). For all the dopants, the content of the metallic element increases with sputtering as observed in as-milled samples (**Annexe D**). This result indicates that they are still located beneath the surface rather than on the surface. The heating step does not change the composition of the surface and of regions close to the surface. A correlation between the quantity of metal M from the dopant and of Al is observed and it can be related to the formation of the Al-M phases observed by XRD (**Figure 3.21**). For the chemical state of the elements, there is also no change for Ti, which is already entirely in the metallic state, and for Sc indicating that no new phases are formed. The absence of change in the chemical state of the dopant during the first desorption step indicates that the reduction of the dopant to metal during milling is a very important step: the best performance in desorption is obtained for $TiCl_3$ that has been completely decomposed.

3.3.2 Catalytic mechanism for the H_2 absorption

From the XPS results, the hypothesis that the decomposition of the dopant is important for the desorption reaction can be validated. For this reaction, $TiCl_3$, entirely decomposed to metallic Ti, is the most efficient dopant. On the contrary for the absorption step, $CeCl_3$ and to a less extent $ScCl_3$ are more efficient than $TiCl_3$ especially for low hydrogen pressures when the driving force of the reaction is smaller. The higher absorption rate may be related to the formation of Sc or Ce hydride that would attract H atoms in the material [89], these hydrides are more stable than TiH_2 (ΔH_f^0= -125, -208 and -217 kJ mol$^{-1}H_2$ for TiH_2, ScH_2 and CeH_2, respectively [66]). The formation of TiH_2 has been found to help the absorption reaction [179] and is observed in the sample milled with pure Ti. Nevertheless, the enthalpies of formation of ScH_2 and CeH_2 are similar but $CeCl_3$ is more efficient than $ScCl_3$ because the re-absorption is possible at lower temperature with $CeCl_3$ both for steps (1.6) and (1.7). Therefore, the stability of the hydride of the metallic element of the dopant is not the only factor playing a role. In addition, the presence of Ce-H or Sc-H compounds is not observed in the samples but the Al_4Ce or Al_xSc_{1-x} phases are identified in the XRD patterns after the first decomposition. These phases may have better properties to split the H_2 molecule and then enhance the H absorption. By DFT calculations, such an effect of Ti-Al and Sc-Al phases has been found to participate in the decomposition of $NaAlH_4$ by forming Ti-Al-H or Sc-Al-H complexes that decrease the H_2 desorption energy [93]. In addition, it has been experimentally shown that the formation of Ti-Al phases plays an important role in the hydrogenation process by comparing the formation of Na_3AlH_6 adding TiF_3 or $Ti(OBu)_4$ [111]. The hydrogenation process is faster using TiF_3 and the X-ray absorption spectroscopy analysis of Ti shows that Ti is reduced to the metallic state with mainly Al as close neighbours, whereas Ti-O pairs are observed in Na_3AlH_6 milled with $Ti(OBu)_4$. By analogy, the Al_4Ce phase may also absorb hydrogen to form Al-Ce-H phases (formation of hydrides observed for $AlCe_3$ [188]) that can act as H traps. Fan et al. [135] have reported that Al_4Ce has a similar effect to promote the absorption of hydrogen by a NaH/Al mixture as $CeCl_3$. This phase, formed upon cycling in $CeCl_3$ doped-$NaAlH_4$, may thus be the real catalytic species. Therefore, Ti-Al, Sc-Al or Ce-Al phases formed could be considered as more efficient than pure metal because their ternary hydrides are less stable than the binary hydrides of the pure metals. It has been shown that Ti-Al intermetallics can absorb hydrogen and release it at lower temperature than TiH_2 [189]. Few data are available on Al-Ce-H phases but Spatz et al. [190] have shown that $Ce(M, Al)_2$ compounds (M=Fe or Mn) can release and uptake hydrogen near room temperature. Such compounds would then be more efficient as H trap than Ti-Al compounds since the H uptake/release occurs at lower temperature.

3.3.3 Proposed catalytic mechanism

The results presented in section 3.2 show the better kinetic properties obtained in desorption using $TiCl_3$ as dopant and in absorption using $CeCl_3$. The dopant distribution or the effect on the bonding strength in $NaAlH_4$ could describe the differences between metal and chloride compounds, but it cannot explain the difference between the chloride compounds. Their respective efficiencies are too disparate considering their chemical similarities, indeed. In addition, the distribution of Ti or Ce species within the powder has been found similar by SEM.

The formation of vacancies during milling has been proposed to explain the role of $TiCl_3$ in the reaction mechanisms (see §1.2.1). These vacancies may be formed because of the reduction of the dopant and the formation of NaCl [123, 131, 132]. They create pathways for hydrogen or active species diffusion, e.g. AlH_3. The role of the vacancies in the diffusion has been described using anelastic spectroscopy [128] and the formation of vacancies during decomposition of $NaAlH_4$ has been further evidenced by the changes of the structure during H-D exchanges observed by Raman spectroscopy [120]. These previous studies have compared doped and undoped-$NaAlH_4$. The importance of the formation of vacancies (linked to the formation of NaCl during milling) is illustrated here, though indirectly, by XPS measurements. The complete reduction of the dopant is observed for $TiCl_3$ but not for $ScCl_3$ or $CeCl_3$. The incomplete reduction to the metallic state may be related to the higher thermodynamic stability of $CeCl_3$ and $ScCl_3$ compared to $TiCl_3$ ($\Delta_f H^0$ = -1053.5, -925.1 and -720.9 kJ mol^{-1}, respectively [191]). The best efficiency of $TiCl_3$ may therefore be related to the presence of a larger quantity of vacancies (and of NaCl). The role of vacancies appears to be particularly important for the desorption steps where $TiCl_3$ shows the highest efficiency and the absence of NaCl when doping with pure Ti leads to poor reaction kinetics. The good properties obtained using Ti-nanoclusters [87] may appear contradictory to this scheme. However, the absence of NaCl and of vacancies is compensated in this case by the very small size of the Ti clusters, which have thus a higher catalytic reactivity and are more efficiently dispersed within the material. This size effect is lost during further heating [8] because of the growth of the clusters during cycling. The absence of vacancies leads then to the lower reactions rate observed in the following cycles. The slightly lower reaction rate obtained when doping with $ScCl_3$ or $CeCl_3$ may be related to the lower amount of Na vacancies created during milling because of the incomplete reduction of these dopants.

Nevertheless, a similar efficiency has been found for $ScCl_3$ and $CeCl_3$ for decomposition, while the pressure drops observed during milling suggest that $ScCl_3$ is almost completely decomposed. Another role of the dopant is then expected. It has been reported that NaCl acts as a grain refiner for $NaAlH_4$ or NaH, and Al_3Ti for the Al phase [131]. The larger amount of NaCl after milling with $TiCl_3$ may therefore ensure a better preservation of the small $NaAlH_4$ grains and then better kinetics. This hypothesis is in agreement with the smaller grain and particle sizes observed for the sample milled with $TiCl_3$. The Al_4Ce phase observed after decomposition for the sample milled with $CeCl_3$ has the same crystal structure as Al_3Ti (tetragonal *I4/mmm*) with larger cell parameters. It is assumed that Al_4Ce can also act as a refiner for Al. Nevertheless, the efficiency is lower because it is present in smaller quantity (still presence of $CeCl_3$) and the cell parameters are larger. The presence of a higher amount of vacancies in the sample milled with $ScCl_3$ (more decomposed during milling) compared to the sample milled with $CeCl_3$, may be shadowed by the absence of a grain refiner (only Al_xSc_{1-x} phases detected) in the material.

On the contrary, the presence of vacancies does not appear to be necessary for absorption since the best dopant is $CeCl_3$, which is the less decomposed dopant after milling. The catalytic mechanism for absorption may then be more related to the presence of other phases like Al_4Ce that enables a better splitting of the H_2 molecule and/or attract more efficiently H in the bulk. The $CeCl_3$ phase itself still present in the sample may also play an important role and ensure the good cycling properties obtained with this sample [134]. Therefore, the relative efficiency of the Al-Ti/Sc/Ce phases for acting as H traps may explain the difference observed in the H_2 absorption rate. Considering that these phases are formed during the first desorption stage upon heating and not during milling, they cannot help the synthesis process where $TiCl_3$ is the most efficient dopant; the formation of vacancies may then be the limiting step for this near room temperature process as for desorption.

4. New high capacity complex hydride: $Ca(BH_4)_2$

The metal borohydrides are widely investigated for solid state hydrogen storage because of their very high hydrogen capacity. $Ca(BH_4)_2$ is particularly interesting because, in addition to a high H_2 capacity, suitable thermodynamic properties are predicted with a reaction enthalpy similar to $NaAlH_4$. $Ca(BH_4)_2$ is considered for hydrogen storage only for a few years and a lot of information about its properties is still lacking. In this chapter, results on the H_2 sorption properties of this compound are shown. The first task was to establish a good method for the synthesis of $Ca(BH_4)_2$ what was demonstrated to be rather difficult. All the experience gained on reactive ball milling by studying $NaAlH_4$ is applied for the synthesis of this new compound by this technique. In addition, the reaction kinetics of borohydride are sluggish. So, looking for a good additive (dopant) is a very important issue. For $Ca(BH_4)_2$, the term "additive" will be preferred to "dopant" because it is more neutral considering that a positive effect of the added compounds on the kinetics is not always found (see §1.2.2). The performance obtained using different additives is compared for the synthesis process and H_2 sorption properties. In addition, the reaction mechanisms of hydrogen sorption are investigated as they are still not clarified for $Ca(BH_4)_2$ (**Annexe C2**).

4.1 Synthesis by reactive ball milling

In the previous chapter, reactive ball milling under H_2 atmosphere is shown to be a fast and efficient method for the synthesis of doped-$NaAlH_4$.

The same method was applied to the synthesis of $Ca(BH_4)_2$ through reaction (4.1).

$$2CaH_2 + CaB_6 + 10H_2 \rightarrow 3Ca(BH_4)_2 \qquad \textbf{(4.1)}$$

This new synthesis method for $Ca(BH_4)_2$ has many advantages such as limiting/suppressing by-products and operating under moderate conditions (pressure and temperature, see §1.2.2). An additive is used considering that it has been shown that the absorption of H_2 by a mixture of CaH_2 and CaB_6 is only possible by using an additive [155, 161]. The synthesis of $Ca(BH_4)_2$ by reactive ball milling has not been reported yet: therefore knowledge gained in the previous work on $NaAlH_4$ was used as a starting point.

4.1.1 Milling process

The optimised parameters of milling 4 (**Table 2.1**) for doped-$NaAlH_4$ were directly applied: 100 bar H_2, 37 balls and 550 rpm rotational speed but with the starting mixture 2 CaH_2 + CaB_6 + 0.04 $TiCl_3$ (milling 1 in **Table 2. 2**). $TiCl_3$ was chosen because of its high efficiency for $NaAlH_4$ and also because of its beneficial effect on the decomposition reaction rate for $Ca(BH_4)_2$ [155, 162]. The evolution of pressure and temperature during milling for this first attempt is given in **Figure 4.1a**. As for milling NaH and Al, an initial increase of the temperature is observed at the beginning of the process because of the friction between the balls and the vial walls involving an increase of the pressure. The milling process was interrupted twice to evaluate the efficiency of the process.

From the evolution of the pressure p and the temperature T recorded during milling and using the ideal gas law, the H_2 content in the vial is determined as before for $NaAlH_4$ (**Figure 4.1b**). The variation of the H_2 content at the beginning of the process is probably overestimated because of the overlapping with the effect of the friction between walls and balls that increase T and p. Nevertheless, the quantity of hydrogen is decreasing during the milling process (except artefacts related of the interruption and restart of milling) but at a much slower reaction rate compared to the synthesis of $NaAlH_4$. The time necessary to reach the ignition temperature is longer, ca. 2 h vs. 1 h for $NaAlH_4$, and a higher temperature is necessary, 54°C measured in the lid vs. 45°C for $NaAlH_4$ milled with $TiCl_3$. The decrease of pressure is very slow and the pressure stabilises after 20 h milling without having reached the completion of the reaction.

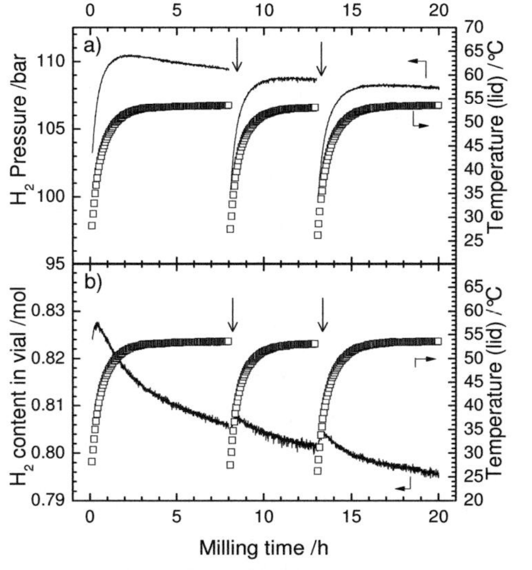

Figure 4.1: Variations of a) pressure (solid line) and temperature (squares) and b) the amount of H_2 in the vial (solid line) and the temperature (squares) for the first milling attempt with the mixture 2 CaH_2 + CaB_6 + 0.04 $TiCl_3$. Downward arrows indicate where the milling was interrupted.

The slow H_2 absorption rate is not surprising because of the sluggish reaction kinetics of borohydride compounds that have already been reported (see §1.2.2). This is partly related to the inertness of CaB_6. The measured decrease of the H_2 content in the vial (after cooling to room temperature) corresponds to a 14% reaction yield considering reaction (4.1) (see **Table 4.1**), whereas more than 95% was obtained after 5 h for the synthesis of $NaAlH_4$. The same strategy as for $NaAlH_4$ was then used to improve the efficiency of the synthesis, namely increasing the energy input and the driving force of the reaction and/or adjusting the amount of additive (see § 3.1.3.1). Therefore, modified conditions were applied: ball-to-powder ratio of 75:1 (increased energy input), 140 bar hydrogen pressure (increased driving force) and a higher quantity of additive with the mixture 2 CaH_2 + CaB_6 + 0.08 $TiCl_3$ (milling 2). In addition, TiF_3 (milling 3) instead of $TiCl_3$ was also used because it has a positive effect during the decomposition of $Ca(BH_4)_2$ [161] and an enhancement for the reverse reaction (synthesis) is thus expected. The different milling parameters are summarised in **Table 2.** 2. The evolution of the temperature and the H_2 content in the vial for these two additional attempts are given in **Figure 4.2**.

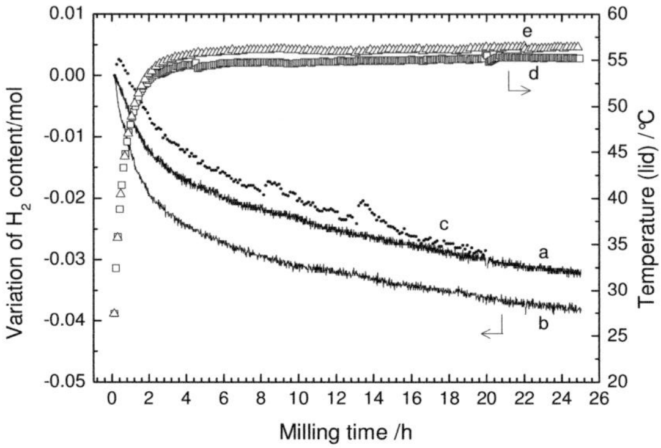

Figure 4.2: Variation of the quantity of H_2 and of the temperature in the vial during the milling of 2 CaH_2 + CaB_6 + 0.08 $TiCl_3$ (a: H_2 content and d: temperature) or + 0.08 TiF_3 (b: H_2 content and e: temperature). The variations observed during the first milling (Figure 4.1b) are given in c for comparison.

A maximum temperature of 55°C (milling 2) is measured in the vial lid and is reached after 5 h milling that is very close to the temperature measured during the first attempt. Assuming that the effect of friction is the same for milling 2 and 3, a direct comparison of the milling conditions used is possible from **Figure 4.2**. The H_2 absorption rate is slightly higher with the modified milling conditions and the total H_2 amount consumed corresponds to a reaction yield of 19 and 28% using $TiCl_3$ and TiF_3 additives, respectively (see **Table 4.1**), and considering reaction (4.1).

Table 4.1: Summary of the variations in H_2 quantity (Δn) after cooling the vial for the different starting mixtures used (calculations using ideal gas law)

Starting mixture	*2 CaH_2 + CaB_6 + 0.04 $TiCl_3$*	*2 CaH_2 + CaB_6 + 0.08 $TiCl_3$*	*2 CaH_2 + CaB_6 + 0.08 TiF_3*
Δn measured (mol H_2)	-0.02194	-0.01861	-0.02792
Δn expected for reaction (4.1) (mol H_2)	-0.153636	-0.099216	-0.101236
Yield (%)	14	19	28

Although small, the yields are quite encouraging considering that the reaction takes place at low temperature (60% was previously obtained at 700 bar and 440°C, after 48 h [155]). It appears also that the use of TiF_3 as additive is slightly more efficient than $TiCl_3$. Nevertheless, the reaction yield is calculated assuming that the hydrogen absorbed has reacted to form $Ca(BH_4)_2$ and not any intermediate phase. This assumption needs to be confirmed and the samples were analysed by XRD to identify the hydride phase formed during milling and/or to evidence the possible presence of new compounds.

The XRD patterns of the milled powders are very similar whatever the additive used and show mainly peaks related to CaH_2 and CaB_6 (**Figure 4.3**), in agreement with the fact that the synthesis of $Ca(BH_4)_2$ is not completed during milling. The peaks related to CaH_2 are very small, likely because of a large broadening of the XRD peaks resulting from the grain size decrease during milling. It is not possible to observe any new phase, in particular also not for the sample milled with TiF_3 that shows the highest reaction yield. The hydride phase formed during milling is likely very disordered or poorly crystallised and cannot be detected by XRD. On the other hand, a small peak is observed that is ascribed to Fe indicating a low contamination of the powder by iron from the vial because of the hardness of CaB_6 that abrades the vial walls.

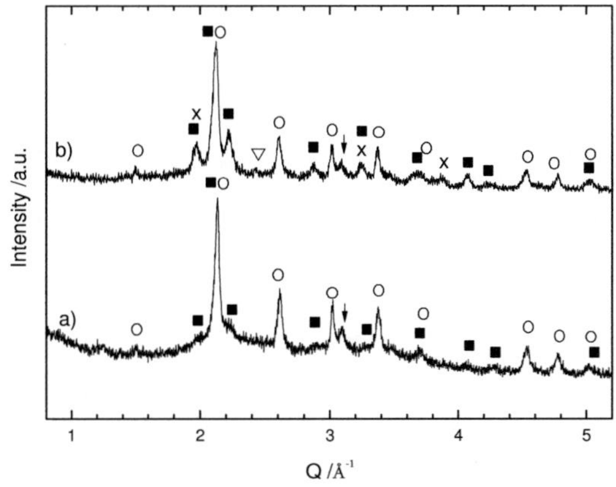

Figure 4.3: XRD patterns for as-milled the mixture 2 CaH_2 + CaB_6 + 0.08 additive: a) $TiCl_3$ and b) TiF_3. The phases present are: CaH_2 (■), CaB_6 (O), CaF_2 (X), TiH_2 (▽) and small traces of Fe (↓)

In addition, no trace of the additive $TiCl_3$ or TiF_3 is found in the patterns and there is also no compound found that could originate from the decomposition of the additive $TiCl_3$. A small amount of CaF_2 and TiH_2 is detected from the XRD pattern of the as-milled sample with TiF_3 that likely originates from the reaction between CaH_2 and TiF_3 following the reaction [192]:

$$3CaH_2 + 2TiF_3 \rightarrow 3CaF_2 + 2TiH_2 + H_2 \qquad \textbf{(4.2)}$$

This reaction releases some hydrogen and thus leads to an underestimated reaction yield as observed for $NaAlH_4$. However, the quantity of hydrogen necessary for the formation of $Ca(BH_4)_2$ is much higher than the H_2 release by decomposing TiF_3 and the reaction yield is only slightly underestimated. In general, the additive itself or any compound formed from it (Ti-based compounds) should be finely distributed in the powder (nanocomposite) or with a highly disordered structure, so hardly detectable by XRD.

4.1.2 Identification of the phases formed during milling

The borohydride compounds are often difficult to analyse by XRD because of their complex structure and sometimes disordered/amorphous state, especially after milling [193]. Therefore, other characterisation techniques must be used. The following discussion is based on the samples after millings 2 and 3 (0.08 $TiCl_3$ or TiF_3) as they show the highest quantity of H_2 absorbed.

4.1.2.1. In-situ XRD

A first indirect proof of the presence of $Ca(BH_4)_2$ in the as-milled sample is given by in-situ XRD measurements performed during heating to 300°C (before decomposition). The patterns obtained at different temperatures are given in **Figure 4.4**. At low temperatures, the patterns are similar to those obtained after milling with the presence of CaH_2 and CaB_6 mainly (**Figure 4.3**). For the sample milled with $TiCl_3$ (**Figure 4.4a**), small peaks arise corresponding to β-$Ca(BH_4)_2$ (high-temperature phase) at 240°C. For the sample milled with TiF_3, these peaks arise at 210°C (**Figure 4.4b**). The peaks of the β phase are growing when the temperature increases.

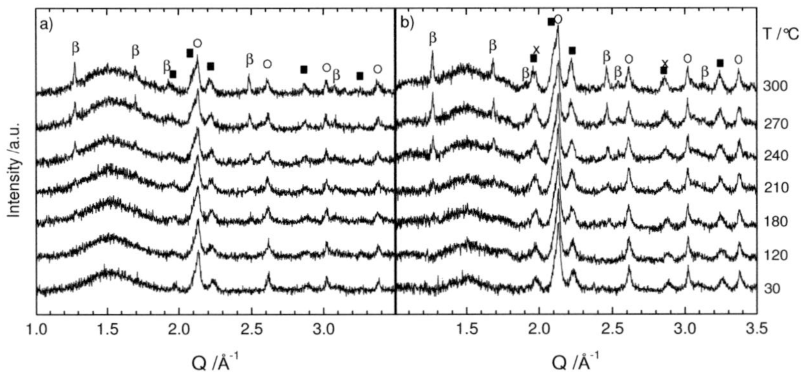

Figure 4.4: in-situ XRD measurements for the as-milled samples with *a)* $TiCl_3$ and *b)* TiF_3. The phases present are CaH_2 (■), CaB_6 (O), CaF_2 (X) and β-$Ca(BH_4)_2$ (β). The hump around 1.5 $Å^{-1}$ corresponds to the diffraction of the capillary.

The appearance of the β-phase is in agreement with previous reports on phase transformations of $Ca(BH_4)_2$ that show the transformation of the room temperature polymorphs α or γ into the β phase above 200 and 270°C, respectively [150]. The growth of the β-$Ca(BH_4)_2$ phase in the milled sample above 200°C is therefore not surprising and confirms that a $Ca(BH_4)_2$ phase is already present in the milled sample. The presence of an intermediate phase in the milled sample, which transforms to $Ca(BH_4)_2$, is unlikely because the samples were heated under Ar and not under H_2 in the capillary. To form $Ca(BH_4)_2$ from any intermediate phase, the presence of H_2 is required (and probably a higher temperature). To confirm directly the presence of $Ca(BH_4)_2$ in the as-milled samples, vibrational spectroscopy techniques, which are not sensitive to the long range order of the material, were used.

4.1.2.2. Vibrational spectroscopy

Raman spectroscopy is often used as an additional characterisation technique for complex hydrides [194-196], especially to detect the presence of borohydride phases while they cannot be identified in XRD patterns [35, 155]. It is also possible to distinguish the different polymorphs of $Ca(BH_4)_2$ by Raman spectroscopy [147]. Raman and IR spectroscopy are sensitive to the vibration of the molecules and show comparable spectra. However, the principles and selection rules are different for both techniques: therefore they give complementary information for a sample (see §2.2.4).

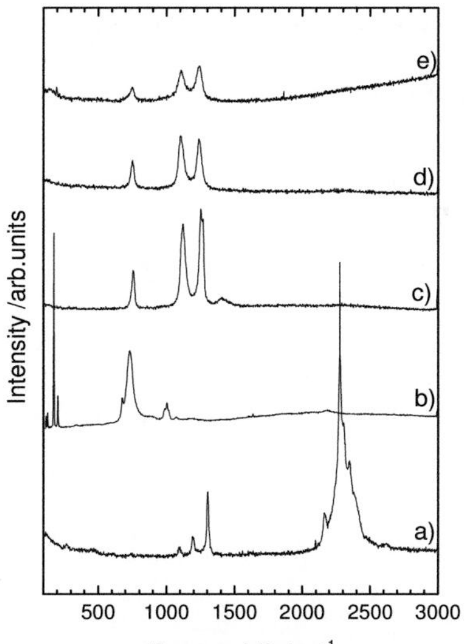

Figure 4.5: Raman spectra measured for the reference compounds a) $Ca(BH_4)_2$, b) CaH_2, c) CaB_6 and for the as-milled samples with d) $TiCl_3$ or e) TiF_3

The Raman spectrum of pure $Ca(BH_4)_2$ (**Figure 4.5a)** is composed of two sets of vibration lines: around 2300 cm^{-1} corresponding to the stretching modes of the B-H bond in $[BH_4]^-$ and in the 1000-1400 cm^{-1} range for the corresponding bending modes [147].

For CaH_2, different lines arise at low wave numbers [197] : 130, 180, 206 cm^{-1}. Additional lines at 727 and 1000 cm^{-1} are present in **Figure 4.5b**. For pure CaB_6, the spectrum [155, 198] is composed of three lines at 755, 1117 and 1259 cm^{-1} (**Figure 4.5c**). The Raman spectra obtained for the samples prepared with $TiCl_3$ or TiF_3 (**Figure 4.5d and e)** show mainly the presence of CaB_6 with the three vibration lines observed for the pure compound. No vibration line corresponding to CaH_2 is observed for the sample containing $TiCl_3$, while there might be a small contribution from CaH_2 in the spectrum of the sample milled with TiF_3 (see at ca. 150 cm^{-1}). There is also no trace of another vibration mode and it is not possible to identify the hydride phase formed during milling. The absence of vibrations detected may be related to the presence of a nanocrystalline structure or/and of metallic elements from the decomposition of the additives that would weaken the Raman signal.

Infrared spectroscopy has a higher cross section than Raman spectroscopy and is thus better for the detection of compounds in small amounts. The results of IR measurements are given in **Figure 4.6.** The IR spectrum obtained for pure $Ca(BH_4)_2$ (**Figure 4.6a**) can be compared with the Raman spectrum in **Figure 4.5a.** There are two main regions of vibrations, around 2300 cm^{-1} for the stretching modes and at 1000-1400 cm^{-1} for the bending modes in the $[BH_4]$ units. The peak observed at *ca.* 1600 cm^{-1} is related to a contamination by water.

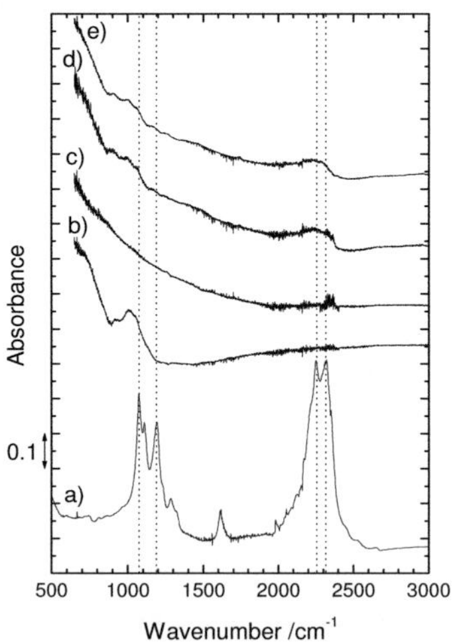

Figure 4.6: IR spectra for a) $Ca(BH_4)_2$, b) CaH_2, c) CaB_6 and the as-milled samples with d) $TiCl_3$ or e) TiF_3.

On the other hand, the spectrum obtained for CaH_2 (**Figure 4.6b**) is different from the Raman spectrum with only a large feature around 1000 cm^{-1}. The spectrum of pure CaB_6 (**Figure 4.6c**) does not show any vibration line in contrast to Raman spectroscopy (the bands observed around 2330-2370 cm^{-1} are related to gaseous CO_2 contamination). The IR spectra obtained for the two as-milled samples are very similar to each other (**Figure 4.6d and e**) with two broad components around 1000 and 2200 cm^{-1}. The component observed around 1000 cm^{-1} is ascribed to CaH_2 and the broad peak around 2200 cm^{-1} may correspond to the vibration in the stretching mode for the B-H bond in $[BH_4]^-$, which is observed in the spectrum of pure $Ca(BH_4)_2$. In the 1100-1200 cm^{-1} range, smaller features are observed that may be the bending mode of B-H besides a CaH_2 contribution.

Considering the small amount of borohydride expected from the quantity of hydrogen absorbed during milling and the low crystallinity, it is expected that the detected signal is very low and not very well defined. The very broad peaks at ca. 2200 cm^{-1} are typical for borohydride compounds and are not present in the case of the intermediate phases already proposed [41, 157, 160]. However, the lines prove the presence of $[BH_4]$ units and the formation of a borohydride phase, not directly $Ca(BH_4)_2$. For example, the formation of $Ti(BH_4)_3$ by a reaction with the additive cannot be excluded (presence of CaF_2 showing the decomposition of TiF_3), as shown by Fang et al. [199]. However, this compound is very unstable and would have decomposed immediately during milling. In addition, the formation of $Ti(BH_4)_3$ reported by Fang et al. occurs only for a reaction between a borohydride and a Ti-based compound, not from the "elements". The combination of different characterisation techniques gives thus a good indication that a part of the milled samples is $Ca(BH_4)_2$. Nevertheless, the wavenumbers of the vibrations for the stretching modes are independent of the $Ca(BH_4)_2$ polymorphs, only the shape is changing. It is therefore not possible to determine which structure was formed during milling because of the small signal.

4.1.2.3. TEM and EELS

Local information on the sample structures and composition were obtained using TEM and EELS analysis. The following description is focused on the analysis of the sample milled with TiF_3. Similar results were obtained for the sample milled with $TiCl_3$ and are not discussed in the following section. Low magnification analysis shows that the particles obtained after milling are large (>1µm) as observed by SEM for $NaAlH_4$ (**Figure 4.7a**). The corresponding EDX analysis (**Figure 4.7b**) shows that all the starting elements Ca, B, Ti and F are present in the particle. Similar EDX spectra as presented in **Figure 4.7b** were obtained on different particles showing a good homogeneity of the sample and indicating a good efficiency of the milling to mix intimately the different compounds. It is also possible to detect the presence of iron from contamination during the milling process and of copper and carbon from the grid used as sample holder.

The analysis at low magnification shows regions in the particles with different contrast indicating the presence of several phases as seen from the structural analysis by XRD or spectroscopy. A typical electron diffraction pattern (**Figure 4.7d**), corresponding to the area indicated by the red square in **Figure 4.7c,** shows the presence of spots corresponding to the electron diffraction of CaB_6 as well as diffuse rings that could be associated to the presence of CaH_2 (see the complete indexation in **Annexe E1**). These rings were observed in the electron diffraction pattern of pure CaH_2 (not shown) but partly correspond to CaO appearing in the sample because of the high reactivity of CaH_2 under the electron beam. The discrete spots corresponding to CaB_6 were also observed for the reference compound (not shown).

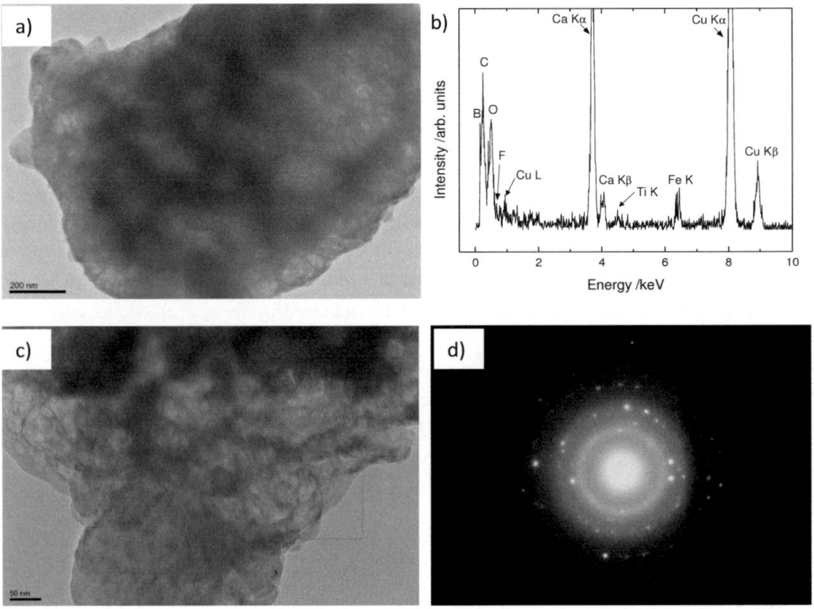

Figure 4.7: TEM analysis of the as-milled sample 2 CaH_2 + CaB_6 + 0.08 TiF_3: a) general view of a particle, b) EDX analysis corresponding to a), c) high magnification view of another particle and d) electron diffraction pattern of the area specified in c).

TEM analysis at higher magnification allowed the study of the local structure in different regions where various properties/structure can be differentiated. The analysis of the area shown in **Figure 4.8a** gives additional information on the sample structure and composition. The corresponding electron diffraction pattern (**Figure 4.8b**) shows mainly spots related to CaB_6 (see **Annexe E2**), but a ring close to the centre of the pattern is also observed corresponding to a crystalline structure with large lattice parameters. The interplanar distance derived from the pattern for this ring is 5.6 Å. The closest lattice spacing found is 5.2 Å for the main diffraction peak of α-$Ca(BH_4)_2$. Values expected for the starting compounds CaH_2 and CaB_6, or for CaF_2, as well as for the intermediate phases proposed for the decomposition of $Ca(BH_4)_2$ [157, 160] are much different from the observed 5.6 Å distance. Nevertheless, the identification of this diffracting phase (ring) as α-$Ca(BH_4)_2$ is questionable because of the difference in the determined distance values. The phase observed in the milled sample could then be related to an intermediate phase not described yet.

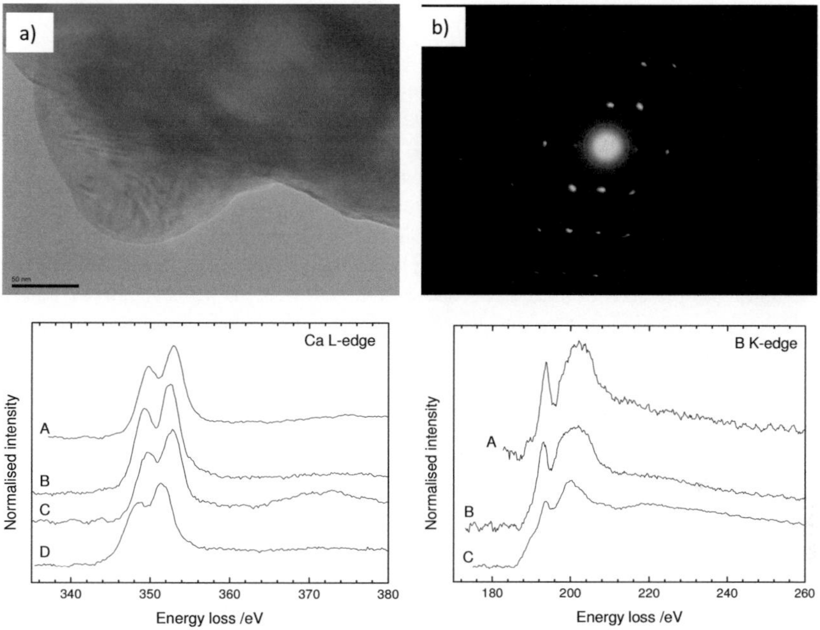

Figure 4.8: a)TEM micrograph of a selected area of the as-milled sample 2 CaH_2 + CaB_6 + 0.08 TiF_3, b) corresponding electron diffraction pattern, c) Ca L edge and d) B K edge EELS spectra (A: as-milled sample, B: $Ca(BH_4)_2$, C: CaB_6 and D: CaH_2).

Additional information is given by the EELS analysis of the Ca L- and B K-edges regions. The Ca L-edges spectra measured for the milled sample (**Figure 4.8c,** curve A) are compared to those of pure $Ca(BH_4)_2$, CaB_6 and CaH_2, (**Figure 4.8c,** curves B, C and D, respectively). The Ca-based compound(s) present in the as-milled sample cannot be obviously determined. The Ca L-edges are very similar for all compounds with a main peak composed of two maxima. The hump at higher energy can be explained by the higher crystallinity of the reference CaB_6 powder compared to other compounds but does not give any information on the sample composition. On the other hand, the shape of the B K-edge obtained for the milled sample (**Figure 4.8d**, curve A) is closer to the shape obtained for pure $Ca(BH_4)_2$ (curve B) than for CaB_6 (curve C).

The different observations made with TEM show the presence of small domains composed of a phase with a signature close to $Ca(BH_4)_2$ but its structure could be slightly different and a CaB_6 phase. Considering that $Ca(BH_4)_2$ is very sensitive to the electron beam of the TEM, the new phase may be a decomposition product of $Ca(BH_4)_2$ prepared by ball milling.

4.1.3 Conclusions on synthesis

Reactive ball milling under high hydrogen pressure allows the synthesis of $Ca(BH_4)_2$ but with a low yield. The combination of different characterisation techniques is necessary for its unambiguous identification, hardly possible by XRD because it is poorly crystallised. The presence of $Ca(BH_4)_2$ can be indirectly determined by the appearance of β-$Ca(BH_4)_2$ observed by XRD during heating that is likely related to the recrystallisation of the disordered phase obtained after milling. The identification of $Ca(BH_4)_2$ in the as-milled sample can be done by IR spectroscopy revealing the presence of vibrations in the $[BH_4]$ unit, whereas Raman spectroscopy only shows the presence of CaB_6. The detection of $Ca(BH_4)_2$ by IR and not by Raman spectroscopy can be explained by the higher cross section of IR allowing the identification of $Ca(BH_4)_2$ present in small quantity. In addition, TEM analysis shows the presence of a new phase with structural properties close to $Ca(BH_4)_2$ that may be a decomposition product of $Ca(BH_4)_2$ produced by the effect of the electron beam.

The presence of $Ca(BH_4)_2$ indicates that the quantity of hydrogen absorbed during milling is actually used for the formation of this phase and the reaction yields can be validated at 19 or 28% using $TiCl_3$ or TiF_3 as additive, respectively. The comparison of the effect of the additives on the milling process shows hardly any difference. It appears that the use of TiF_3 leads to a slightly higher reaction yield than using $TiCl_3$. The yields are quite small and far from the complete synthesis reaction, but it is quite encouraging considering that the synthesis process is done near room temperature. The analysis of the decomposition of the as-milled samples should give more information about $Ca(BH_4)_2$ formed during milling and the effect of the additives.

4.2 Decomposition of $Ca(BH_4)_2$: kinetics and mechanism.

4.2.1 Study of pure $Ca(BH_4)_2$

Before analysing the decomposition behaviour of the milled samples, it is interesting to investigate the decomposition steps for pure as-received $Ca(BH_4)_2$. The decomposition of $Ca(BH_4)_2$ has been shown to proceed in two steps seen by DSC and TG measurements with CaH_2 and CaB_6 as final products following the overall reaction (1.11) [155, 156]. The first step releases ca. 6 wt% H_2 and corresponds to a sharp and very intense peak in the DSC trace while the second step is broader and releases the remaining H_2 for a total of 9.4 wt% H_2 [157] (see §1.2.2). The intermediate phase is not really identified. Riktor et al. [160] have reported the formation of CaB_2H_x found by ex-situ XRD but have indicated that this phase is not observed when the decomposition is analysed by in-situ synchrotron measurements. In addition, the decomposition temperatures for $Ca(BH_4)_2$ reported so far (see §1.2.2.4) are higher than 300°C. These values are much higher than expected from the calculated enthalpy of reaction (1.11).

Pure $Ca(BH_4)_2$ compound was analysed by DSC in different atmospheres to study the decomposition route. **Figure 4.9** gives the DSC traces obtained under Ar and 3 or 7 bar H_2. In all traces, the decomposition of $Ca(BH_4)_2$ occurs as already described in two endothermic steps above 350°C (peak 1 and 2), with the first event being much more pronounced than the second one (see §1.2.2.4) [155, 156]. The onset temperature of the first decomposition step is lower under Ar as expected because the decomposition reaction is related only to kinetics and not to the thermodynamic properties as under H_2 pressure. The onset temperature is only ca. 10°C lower than in 3 bar H_2 and indicates that there are likely strong kinetic limitations for the reaction that control the first decomposition temperature. The poor kinetics observed are nevertheless not surprising because there is no additive and the structure is rather coarse for the as-received compound. By comparing the traces at 3 and 7 bar H_2, an increase of the temperature for peak 1 is observed when increasing the H_2 pressure, in agreement with the van't Hoff equation (2.8), i.e the thermodynamic properties control the first event when measuring in H_2 atmosphere.

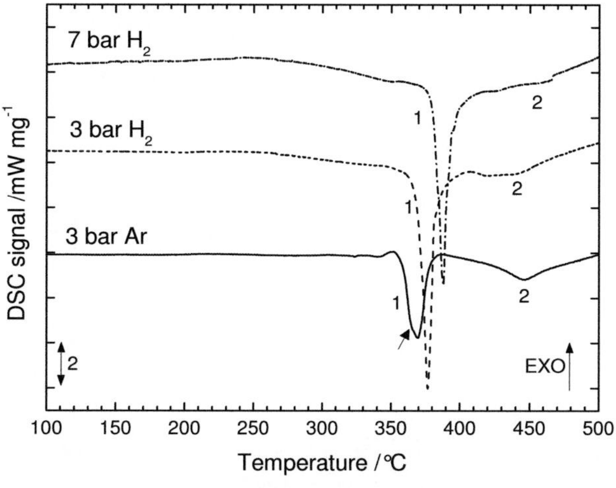

Figure 4.9: DSC traces (5°C min^{-1}) for as-received $Ca(BH_4)_2$ in different atmospheres. The arrow shows the shoulder at low temperature of peak 1 measured in Ar.

The differences between the traces obtained in Ar or H_2 are very small but the decomposition route appears slightly different with a change in the relative intensities of the two events. It also appears that peak 1 measured under Ar is composed of two events with the presence of a shoulder at low temperature (see arrow in **Figure 4.9**). These observations indicate that the decomposition of $Ca(BH_4)_2$ is more complicated than initially reported in literature with different events overlapping for the two peaks observed by DSC. This is consistent with the report from Riktor et al. [160] who have obtained different XRD patterns after heating ex-situ to temperatures corresponding to the maximum of peak 1 and between reactions 1 and 2.

In addition, the second step (peak 2) occurs at a higher temperature in Ar than in H_2 atmosphere and its intensity decreases when the H_2 pressure increases. Considering that the temperature of the second event does not vary much and that after heating the sample has no more a powder morphology but looks like a "foam" with much larger volume, it is assumed that during the second step a phase transformation similar to melting (as seen for other borohydrides) takes place concomitantly with H_2 release.

4.2.2 Decomposition of the as-milled sample

4.2.2.1. DSC

The samples prepared by reactive ball milling contain only a low amount of $Ca(BH_4)_2$ but can be used for the analysis of the decomposition route. In addition, the comparison with $Ca(BH_4)_2$ available commercially (Aldrich) can give an idea on the quality of the phase formed during milling. First, a DSC measurement was performed at 3 bar H_2 (**Figure 4.10**).

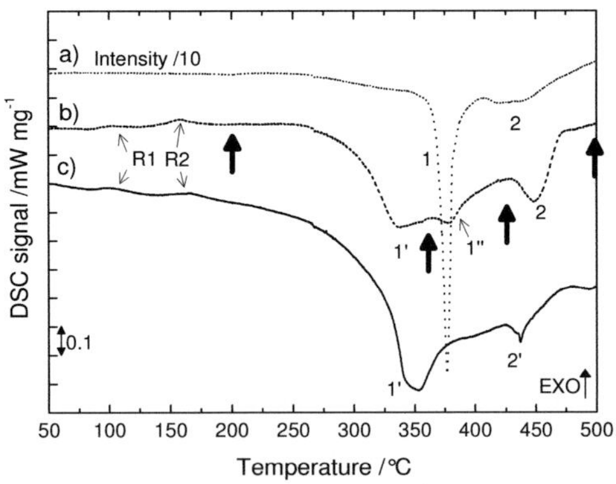

Figure 4.10: DSC traces in 3 bar H_2 (5°C min^{-1}) for *a)* pure $Ca(BH_4)_2$ and as-milled samples with b*)* $TiCl_3$ and *c)* TiF_3 (thick arrows show the temperature chosen for ex-situ XRD analysis, see below)

Between 100 and 200°C, small exothermic peaks (R1 and R2) are detected and correspond to the recrystallisation of the poorly crystallised $Ca(BH_4)_2$ phase obtained after milling as observed by in-situ XRD (**Figure 4.4**). The decomposition (endothermic) of as-received $Ca(BH_4)_2$ occurs at a temperature above 350°C, while the addition of compounds like $TiCl_3$ and TiF_3 decreases the decomposition temperature of the first decomposition step 1' (temperature of peak maximum *ca.* 40°C lower). The decomposition temperature is similar for the samples milled with both additives, however, three reaction steps are observed for the sample milled with TiF_3 whereas only two are distinguished for the sample milled with $TiCl_3$ (as for pure $Ca(BH_4)_2$). The overlapping of two reaction steps for peak 1 is supposed from the analysis of as-received $Ca(BH_4)_2$ (see previous section) and in the sample milled with TiF_3, peak 1 may be separated in two steps 1' and 1" because of a change of the kinetic properties.

The decrease of the decomposition temperature in measurements performed at 3 bar H_2 for samples milled with an additive is an indication that the onset temperature is not only related to thermodynamic properties in H_2 atmosphere but is also linked to the very slow kinetics of the reaction (see §4.2.1), as reported previously for pure $NaAlH_4$ (§3.2.1). This is in agreement with the calculated decomposition enthalpy (see §1.2.2) that predicted a lower decomposition temperature at 3 bar H_2 (ca. 100°C, see **Figure 1.9**). On the opposite, peak 2 at 425-450°C appears to be little affected by the addition of Ti-based additives. This result confirms that this event is partly related to a phase transformation.

4.2.2.2. TDS

Thermal desorption spectroscopy (TDS) measurements in high vacuum were also performed to analyse the gases released during decomposition. The reactor containing the sample is mounted in line of sight of a mass spectrometer; this set-up allows a very precise analysis of the desorbed gases. The results obtained for the as-received $Ca(BH_4)_2$ and CaH_2 and for the as-milled samples are given in **Figure 4.11.** The decomposition proceeds in a slightly different way than observed by DSC performed in H_2 because the experiments are here performed in vacuum.

For as-received $Ca(BH_4)_2$, the decomposition starts with a first small increase of the H_2 pressure (marked with 0) observed before the main desorption peak 1 (maximum at ca. 345°C). Afterwards, a second event, labelled 2, releases H_2 with a maximum at 380°C. The decomposition of as-received CaH_2 starts at a temperature above 500°C (peak 3).

For the as-milled samples, no major difference is observed between the addition of $TiCl_3$ or TiF_3. The first decomposition step (1') starts at ca. 230°C and has a lower intensity than the second step (2') starting at 340°C. The decomposition (peak 1') starts at a lower temperature than for as-received $Ca(BH_4)_2$, because of the presence of additives and in agreement with the DSC results. A third event (3') releases H_2 with a maximum at 520-530°C and is likely related to the decomposition of CaH_2. The release of H_2 occurs at a lower temperature than for pure CaH_2 because of the use of additives and of the milling process that reduces the desorption temperature of metal hydrides. The main difference between the decomposition of the as-milled sample and of as-received $Ca(BH_4)_2$ is the relative intensities of the two main decomposition events 1/1' and 2/2'.

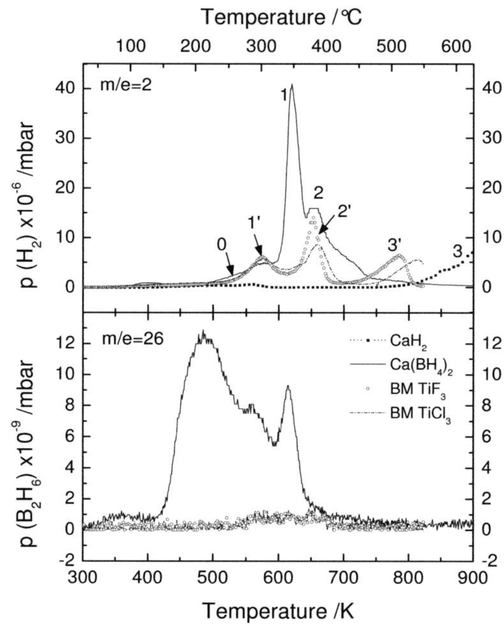

Figure 4.11: TDS spectra of H_2 (top curve) and B_2H_6 (bottom curve) release for as-received CaH_2 and $Ca(BH_4)_2$ and for as-milled samples with TiF_3 (BM TiF_3) and $TiCl_3$ (BM $TiCl_3$).

In addition, the release of B_2H_6 is considered. For as-received $Ca(BH_4)_2$, a significant amount of B_2H_6 is released between 130 and 410°C, apparently in two steps. The first one occurs before and the second one together with the first H_2 release peak 1. The evolution of B_2H_6 may not be totally related to the decomposition of $Ca(BH_4)_2$, in particular at low temperature, but can also originate from the evaporation of solvent residues still present in the powder after wet chemistry synthesis. No significant release of B_2H_6 is observed for the milled samples.

As a summary, the decomposition of as-received $Ca(BH_4)_2$ in vacuum starts by a small release of H_2 and B_2H_6 from 130°C and is followed by two more important decomposition steps at 320-370

(H_2 and B_2H_6) and 370-400°C (only H_2). For the milled samples, smaller intensities of H_2 release are obtained since a small yield of $Ca(BH_4)_2$ is present (same quantity of starting powder in the reactor). The first decomposition step starts before the main event observed for as-received $Ca(BH_4)_2$ and shows that the addition of Ti-based additive improves the kinetic properties of $Ca(BH_4)_2$. In TDS, it is not possible to distinguish the two reactions overlapping in peak 1' for the sample milled with TiF_3. On the other hand, the second step observed for the milled samples arises at the same temperature as for as-received $Ca(BH_4)_2$ in agreement with the DSC measurements. This may be a confirmation that this decomposition step is partly related to a phase transformation event.

4.2.2.3. XRD analysis of the decomposition route

To understand the reaction path for the decomposition of $Ca(BH_4)_2$, the as-milled sample with TiF_3 was heated at 5°C min^{-1} under H_2 atmosphere to different temperatures corresponding to different reaction stages observed by DSC and shown by arrows in **Figure 4.10**. XRD analysis was performed on the differently heat treated samples and the patterns obtained are given in **Figure 4.12**. The first sample was heated to 200°C, i.e. after the occurrence of the small exothermic events R1 and R2 detected by DSC but before the decomposition. The corresponding XRD pattern (**Figure 4.12a**) shows small peaks related to β-$Ca(BH_4)_2$ as also seen by in-situ XRD (**Figure 4.4**). This result confirms that the exothermic events are related to the recrystallisation of the disordered $Ca(BH_4)_2$ phase obtained after milling.

The second sample was heated to 360°C corresponding to the end of the decomposition event 1' (**Figure 4.12b**). The XRD pattern shows the formation of the intermediate phase CaB_2H_x described by Riktor et al. [160]. In addition, the XRD peaks of the CaH_2 phase are more intense, i.e. the quantity of CaH_2 has increased or the peaks are sharper because of grain growth upon heating.

The third sample was heated to 425°C after the endothermic event 1'. The corresponding XRD pattern (**Figure 4.12c**) shows that the intermediate phase CaB_2H_x has almost completely disappeared but no new phase is detected. The peaks related to CaH_2 and CaB_6 remain unchanged. The formation of another intermediate phase is thus assumed with a poorly crystallised or amorphous-like structure, as also reported by Riktor et al. [160]. This result indicates that events 1' and 1" obtained for the sample milled with TiF_3 correspond to two overlapping steps for event 1 obtained for pure $Ca(BH_4)_2$.

After the heat treatment to 500°C, i.e. after the complete decomposition of the sample, the XRD pattern (**Figure 4.12d**) shows the presence of CaH_2, CaB_6 and CaO. It is difficult to detect any change in the peaks related to CaH_2 or CaB_6 which are the decomposition products of reaction **Error! Reference source not found.**). The formation of CaO likely originates from contamination during the XRD measurement. However, this phase is usually not observed in similar samples containing CaH_2 or CaB_6, so elemental Ca has likely been formed rather than

CaH_2 or CaB_6 during the last step of decomposition (of the amorphous-like intermediate phase). This elemental Ca reacts then very fast with oxygen from air contamination occurring during XRD measurements. The decomposition path observed by this experiment is very similar to the one reported by Riktor et al. [160] for a mixture of β- and γ-$Ca(BH_4)_2$.

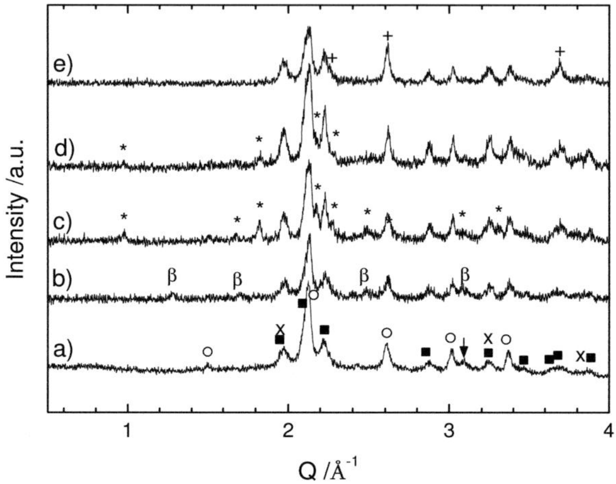

Figure 4.12: XRD patterns for 2 CaH_2 + CaB_6 + 0.08 TiF_3 : a) as-milled and after ex-situ heating (in 1 bar H_2) to b) 200°C, c) 360°C, d) 425°C and e) 500°C. The phases present are: CaH_2 (■), CaB_6 (O), CaF_2 (X), β-$Ca(BH_4)_2$ (β), CaB_2H_x (*) and CaO (+).

4.2.3 Summary of the results on $Ca(BH_4)_2$ decomposition

A very important information from the XRD analysis of the decomposition steps under H_2 for $Ca(BH_4)_2$ prepared by reactive ball milling is that the process is similar to the one described for pure $Ca(BH_4)_2$ prepared by wet chemistry. The addition of Ti-based additives decreases the decomposition temperature to ca. 300°C. A difference observed for the sample milled with TiF_3 is the splitting of the first decomposition step commonly reported for $Ca(BH_4)_2$ in two sub-steps.

The analysis of the decomposition by DSC, TDS and XRD shows that the reactions have very strong kinetic limitations that alter the decomposition route according to the conditions used, in particular the atmosphere. Shim et al [45] have also reported an influence of the atmosphere for a different but comparable system (reactive hydride composite $LiBH_4$ + metal hydride). They have ascribed the better H_2 desorption properties with 3 bar H_2 backpressure to the absence of $Li_2B_{12}H_{12}$ and amorphous boron usually formed during decomposition of $LiBH_4$ in Ar. These products prevent the reaction between $LiBH_4$ and the metal hydride. The formation of different intermediate phases may also explain the difference observed for the DSC traces of the milled sample with TiF_3 or $TiCl_3$ showing a decomposition route in 3 or 2 steps, respectively. Nevertheless, these differences may also be ascribed to kinetic effects with the overlapping of the first two events for the sample milled with $TiCl_3$ (as for as-received $Ca(BH_4)_2$) that explains the similar route observed by TDS measured in vacuum.

From all the results obtained, it appears that the first decomposition step observed for pure $Ca(BH_4)_2$ (around 370°C in Ar) is in reality composed of two reaction steps that can be distinguished by adding TiF_3 The first intermediate phase formed during heating in H_2 is the CaB_2H_x phase described by Riktor et al. [160] but it is not possible to identify the second one, which shows a amorphous-like structure by XRD. For the last reaction step (around 450°C in Ar), this amorphous-like intermediate decomposes and a phase transformation could also occur simultaneously considering that this event is little affected by the atmosphere used or by the additives. The release of H_2 simultaneously to melting is commonly reported for borohydride compounds [200]. Here, the phase transformation should be the melting or a kind of relaxation for the amorphous-like phase or for another intermediate compound. It is likely that several intermediate phases are favoured according to the measurement conditions in relation to their different kinetic and thermodynamic properties. The decomposition of $Ca(BH_4)_2$ appears then much more complicated than described up to now (see § 1.2.2.4).

4.3 Effect of additives on the reversibility

$Ca(BH_4)_2$ prepared by reactive ball milling shows a similar decomposition behaviour as the as-received powder but the reaction yield during milling is rather low. Improving this by changing the milling process is quite difficult because the conditions used are close to the maximum technical specifications of the milling vial. In addition, more information on the role of the additive is needed. For that purpose, the reversibility of the milled samples was tested by hydrogenation at higher temperature. One desorption/absorption cycle was performed in a Sieverts' type apparatus.

4.3.1 Re-hydrogenation of decomposed samples in Sieverts' and DSC apparatus

The samples milled with $TiCl_3$ or TiF_3 (milling 2 and 3) were subjected to one cycle with a first step of desorption during heating to 350°C (static vacuum) and re-absorption at 100 bar H_2 and 350°C. The first desorption step leads to comparable results for both samples milled with $TiCl_3$ or TiF_3 (**Figure 4.13a**) with a release of *ca.* 1.6 wt% of hydrogen. This value is in agreement with the amount of gas absorbed during milling with $TiCl_3$ but lower than absorbed by the sample milled with TiF_3. This is related to a slightly lower rate for the release of H_2. The decomposition started at 270-300°C for both samples with a hydrogen backpressure of about 0.2 bar in agreement with the results obtained by more precise DSC measurements.

The results of the following absorption step are shown in **Figure 4.13b**. For both samples, a first part ($t < 2$ h) with fast hydrogen absorption occurs and then a continuous and slow increase of the hydrogen uptake is observed for the remaining time. This behaviour may indicate the occurrence of two reaction steps. Firstly, the formation of an intermediate phase with a fast absorption rate and in a second step, hydrogenation with a slower rate of this intermediate to form the final product. Nevertheless, no intermediate has been reported so far for hydrogenation. For the sample with TiF_3, a capacity of 5.7 wt% H_2 is achieved after 24 h whereas only 2.8 wt% H_2 is reached in the sample containing $TiCl_3$. This difference is explained by a larger H_2 uptake in the two absorption steps showing the higher efficiency of TiF_3 for re-hydrogenation. The capacity obtained using $TiCl_3$ remains far from the theoretical value of 9.6 wt% H_2 (ca. 9 w% H_2 taking into account the weight of the additive). The maximum capacity obtained for the sample containing TiF_3 is 6.2 wt% H_2 after 50 h hydrogenation (67% reversibility considering the weight of the additive). This value is in agreement with the reversible absorption (60%) of hydrogen obtained by Kim *et al.* [161] under similar conditions (90 bar H_2 and 350°C for 24 h). TiF_3 appears as a much more efficient additive than $TiCl_3$ for re-hydrogenation.

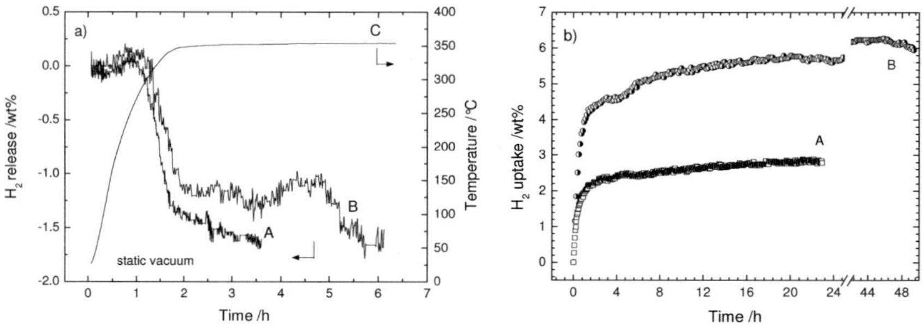

Figure 4.13: a) First desorption in static vacuum for as-milled samples with $TiCl_3$ (A) and TiF_3 (B) during heating to 350°C (temperature curve C) and b) following re-hydrogenation at 100 bar H_2 and 350°C.

The better cycling ability of the sample doped with TiF_3 is also shown by DSC measurements performed under different H_2 pressures (**Figure 4.14**). The first cycle was performed at 3 bar H_2 to decompose the compound(s) formed during milling (curve ①). The DSC traces during heating to 500°C are composed of several endothermic events occurring from ca. 250°C and related to the release of H_2 (see §4.2). No events can be seen in the following cooling stage since the pressure is too low for re-hydrogenation. In this first cycle, no large difference is observed between the samples milled with $TiCl_3$ or TiF_3, except in the number of reaction steps as already described previously. The second cycle (curve ②) is performed at 100 bar H_2 and for the sample milled with TiF_3 a very large exothermic signal is obtained that starts around 150°C and is related to H_2 absorption. On the opposite, a much weaker exothermic event is observed for the sample milled with $TiCl_3$. The area (ΔQ) of the exothermic peak (hatched area) is directly linked to the amount of H_2 absorbed during heating. The larger area of the exothermic peak obtained for the sample milled with TiF_3 shows the higher efficiency of this additive in re-absorption. The third cycle at 3 bar H_2 (curve ③) shows a very broad endothermic DSC signal starting at lower temperature than in the first cycle and is likely related to the decomposition of the hydride phases ($Ca(BH_4)_2$ or intermediate) formed during absorption. However, it is not well defined probably because of the slow reaction rate that is typical of borohydride compounds and that hinders the formation of a well crystallised hydride in the second cycle.

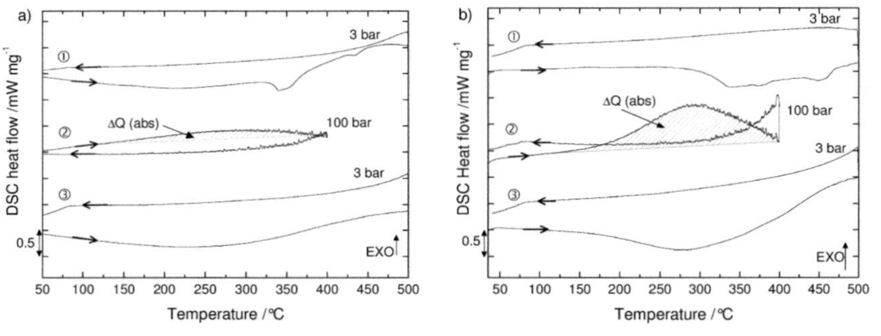

Figure 4.14: DSC traces during cycling of the as milled samples with a) $TiCl_3$ and b) TiF_3 (same scale). Cycles are performed at 5°C min^{-1}, cycle ① at 3 bar H_2, cycle ② at 100 bar H_2 and cycle ③ at 3 bar H_2. The hatched areas correspond to H_2 absorption (exothermic peak).

4.3.2 Analysis of the phases formed after re-hydrogenation

The structural analysis of the re-hydrogenation products is necessary to understand the difference observed between $TiCl_3$ and TiF_3 used as additives. Significantly different phases are observed in the XRD patterns of re-hydrogenated samples depending on the additive used (**Figure 4.15**). For both samples, un-reacted CaH_2 and CaB_6 are observed in agreement with the reaction yield obtained during the re-hydrogenation stage (**Figure 4.13**). In addition, there is the formation of $CaCl_2$ or CaF_2 likely from the reaction of the additive with CaH_2. The peaks indicated by ▲ in **Figure 4.15a,** at positions expected for $CaCl_2$, have very high relative intensity; it is therefore very likely that there is an overlapping with peaks of other compounds, not identified yet. $CaCl_2$ is likely formed from the reaction of CaH_2 and $TiCl_3$, similar to reaction (4.2). The presence of TiH_2 may also be seen in **Figure 4.15a**. For the sample milled with $TiCl_3$, the decomposition of the additive occurs during cycling, whereas the decomposition of TiF_3 is observed already after milling.

For the sample with TiF_3, intense peaks corresponding to the α- and the β-phase of $Ca(BH_4)_2$ appear. On the contrary, in the sample with $TiCl_3$, only a weak peak related to α-$Ca(BH_4)_2$ is obtained. Besides, there is also the formation of another unknown phase (indicated with question marks in **Figure 4.15a**). These new peaks do not correspond to any intermediate phase [157, 160] reported for the decomposition of $Ca(BH_4)_2$ or any polymorph of $Ca(BH_4)_2$. Under other milling conditions (not detailed here) with $TiCl_3$ as additive, the formation of CaHCl is observed after cycling. Therefore, the un-indexed XRD peaks may be related to a similar Ca-H-Cl phase as suggested by Kim *et al.* [162] after ball milling $Ca(BH_4)_2$. A rough comparison of the XRD patterns obtained by these authors with those obtained in this study (**Figure 4.15**) shows some similarities in the position of the unindexed peaks. The formation of such a phase is consistent with the weak peaks of CaH_2 and limits the formation of $Ca(BH_4)_2$ during absorption. No intermediate phase for the formation of $Ca(BH_4)_2$ is detected in the sample with TiF_3 even though the reaction is not complete. From the XRD pattern, it cannot be determined if Ti, produced by the decomposition of the additives, forms a new compound or remains metallic.

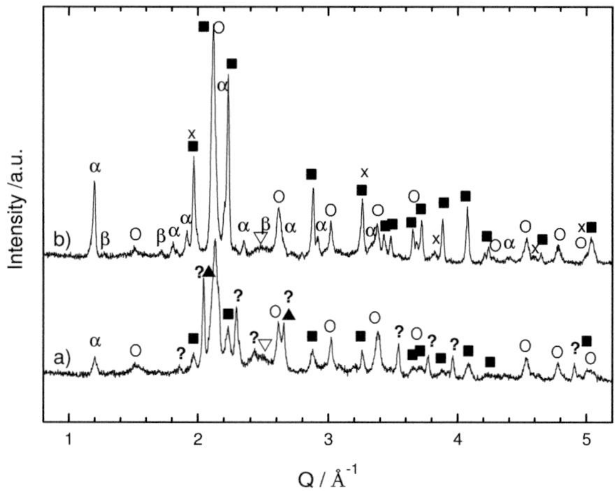

Figure 4.15: XRD patterns for re-hydrogenated samples milled with a) $TiCl_3$ and b) TiF_3. The phases present are: CaH_2 (■), CaB_6 (O), CaF_2 (X), $CaCl_2$ (▲), α-$Ca(BH_4)_2$ (α), β-$Ca(BH_4)_2$ (β) and TiH_2 (▽). Peaks marked with (?) could not be indexed.

IR spectroscopy is performed to check for phases not detected by XRD (**Figure 4.16**). The IR spectrum of the sample with $TiCl_3$ shows mainly a weak feature related to CaH_2. On the contrary, $Ca(BH_4)_2$ vibration lines appear clearly for the sample with TiF_3 with a similar shape as the α phase in agreement with XRD. This is a confirmation that after cycling, no or little $Ca(BH_4)_2$ is formed in the sample with $TiCl_3$. The phase formed in this case may be a Ca-H-Cl phase as already supposed from the XRD measurement: the small peaks around 1200cm^{-1} observed in the IR spectrum (see arrow in **Figure 4.16d**) might be its IR sign.

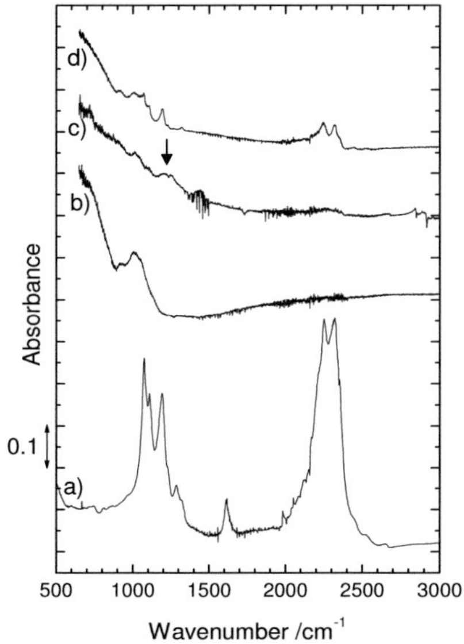

Figure 4.16: IR spectra of a) $Ca(BH_4)_2$, b) CaH_2, and of the re-hydrogenated samples prepared with c) $TiCl_3$ and d) TiF_3.

DSC measurements were also performed for the re-hydrogenated sample after one desorption and absorption cycle. The DSC traces obtained for the sample as-milled with TiF_3 and after re-hydrogenation are given in **Figure 4.17** and compared with the DSC trace of pure $Ca(BH_4)_2$. The trace of the re-hydrogenated sample (curve c) is close to the one obtained for pure $Ca(BH_4)_2$ but with much lower intensity, confirming the re-formation of $Ca(BH_4)_2$ after a complete cycle. There is a small endothermic peak (P) arising at 145°C followed by an intense endothermic peak starting at ca. 300°C (1') and a smaller one (2'), which is not very well defined, around 450°C. The formation of the α phase is observed after cycling by XRD and IR spectroscopy and the peak P is related to the phase transformation from the α to the β phase of $Ca(BH_4)_2$. The two following events (1' and 2') are related to the decomposition of $Ca(BH_4)_2$ in two steps as for the pure compound but the first decomposition step starts at lower temperature as for the as-milled sample. $Ca(BH_4)_2$ samples after milling and after re-hydrogenation have slightly different decomposition routes because it is not possible to observe the two sub-steps 1' and 1" of the first decomposition step for the re-hydrogenated sample. It is possible that $Ca(BH_4)_2$ formed by re-hydrogenation at high temperature favours an overlapping of these two sub-steps whereas it is possible to distinguish them in the as-milled phase because of a poorly crystallised structure or a modification of the additive effect.

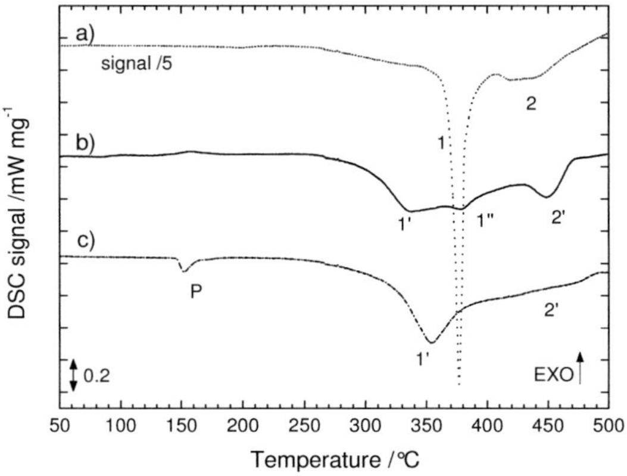

Figure 4.17: DSC traces (5 K min^{-1}) at 3 bar H_2 for a) pure $Ca(BH_4)_2$ (DSC signal intensity divided by 5), b) as-milled and c) re-hydrogenated sample with TiF_3.

4.3.3 Discussion on the higher efficiency of TiF_3 for re-hydrogenation

From the analysis of the re-hydrogenation of decomposed $Ca(BH_4)_2$, a difference in efficiency is shown for $TiCl_3$ and TiF_3, with the formation of a higher amount of $Ca(BH_4)_2$ using TiF_3 as additive. Such an observation of a different behaviour for re-absorption using different additives has not been reported so far. The formation of an unknown phase when using $TiCl_3$ as additive likely prevents the formation of $Ca(BH_4)_2$. In contrast, with TiF_3, $Ca(BH_4)_2$ is formed during re-hydrogenation with no by-products that limit reaction (4.1). The formation of the unknown phase (ascribed to Ca-H-Cl) may be explained by a higher reactivity of $TiCl_3$ towards borohydride compounds, as observed by Fang *et al.* [199] in the case of $LiBH_4$ (higher driving force for the reaction between $LiBH_4$ and $TiCl_3$ vs. TiF_3). A similar behaviour between $Ca(BH_4)_2$ and $TiCl_3$ explains then the formation of the unknown compound Ca-H-Cl. The absence of a by-product is a clear advantage of TiF_3 but the exact influence of this compound is unclear.

Little information is available in the literature on the sorption reactions and doping in $Ca(BH_4)_2$. Different possible mechanisms are proposed from the experimental results depending on the influence of Ti-based (e.g. TiH_2, TiB_2) or fluoride (e.g. CaF_2, $CaH_{2-x}F_x$) compounds formed by the (partial) decomposition of TiF_3 observed here and/or the effect of the F anion. From a thermodynamic point of view, F may substitute H in $Ca(BH_4)_2$ and change the enthalpy of the absorption reaction. It has been demonstrated for Na_3AlH_6 by calculation and verified with experiments [83, 201] that the fluorine anion can substitute H in the structure. This substitution changes the thermodynamic properties, *e.g.* by changing the enthalpy of reaction, but at the expense of the H_2 capacity. It has been predicted by density functional theory that the same concept is applicable to $LiBH_4$ [202]. From the characterisation of the milled and cycled samples, it is not possible to know whether some H atoms are substituted by F in $Ca(BH_4)_2$ but the same concept may be applied to this compound.

Considering the improvement of the kinetic properties, the catalytic activity of Ti species (TiH_2, metallic Ti or Ti-based alloys formed during milling/heating), already demonstrated for sodium alanate [114], can also promote the H_2 absorption in the CaH_2 and CaB_6 mixture. The Ti species are formed by the decomposition of the Ti-based additive. TiF_3 is more easily decomposed than $TiCl_3$ because of the higher reactivity of F towards CaH_2 to form CaF_2 (observed after milling), explained by the "easy" substitution of H atoms by F atoms [203]. This is shown by the presence of CaF_2 already after milling, whereas $CaCl_2$ appears only after cycling at 350°C. A higher quantity of catalytic Ti species may then be formed in this sample explaining the better efficiency of the hydrogen absorption. From the XRD analysis, the presence of TiH_2 is detected. This compound is known to improve the kinetic properties of H_2 sorption for metal hydrides, although it is not the most efficient catalyst [192]. In the case of MgH_2 mixed with TiF_3 or $TiCl_3$, Ma *et al.* [192] have demonstrated that TiH_2 is not the main active species because higher reaction rates are observed when adding TiF_3, while TiH_2 is present after milling with either TiF_3 or $TiCl_3$. On the other hand, all Ti formed after the decomposition of TiF_3 may not form TiH_2. By analogy with the formation of ZrB_2 or TiB_2 reported after milling/cycling a $LiBH_4$ + MgH_2 composite using $ZrCl_4$ [204] or Ti-isopropoxide [39] as additive, respectively, the formation of TiB_2 in the milled and/or cycled samples is proposed. It has been demonstrated that TiB_2 decreases the decomposition temperature of MgH_2 and has a positive effect on the hydrogen desorption at the near surface of MgH_2 [205]. If this effect is not directly applicable to complex hydrides, the reaction between Ti and B (from $Ca(BH_4)_2$ or CaB_6) may weaken the B-H or B-B bonds and favour the reaction rates. It can also act as a grain refiner [39]. In addition, the role of $CaH_{2-x}F_x$ phase(s) (CaF_2 here), formed during milling and cycling, may affect H_2 absorption, as reported by Kim *et al.* [161]. From these phases, Ti-F-Ca compounds may also be formed, similarly to the Ti-F-Mg phases detected by XPS for MgH_2 milled with TiF_3 [192]. The influence of this phase is related to a modification of the electronic structure of Ti, leading for example to a modification of its catalytic activity for the dissociation/ recombination of H_2. The role of different fluoride phases and the possible formation of Ti-based compounds like TiB_2 is a key point to understand the improvement of the H_2 sorption properties. Further investigations using different analytical methods (XAS, XPS, in-situ techniques…) are needed to establish the catalytic and reaction mechanism for $Ca(BH_4)_2$.

4.3.4 Conclusion on the H_2 sorption properties of $Ca(BH_4)_2$

In the previous section, similar decomposition routes are shown for $Ca(BH_4)_2$ prepared by reactive ball milling and as-received $Ca(BH_4)_2$ prepared by wet-chemistry. The decomposition occurs in several steps that overlap during TDS or DSC measurements. The as-milled samples with $TiCl_3$ or TiF_3 show a similar decomposition temperature, lower than for pure $Ca(BH_4)_2$, but more reaction steps are distinguished for the sample milled with TiF_3. It has already been shown that the re-hydrogenation is not possible without the presence of an additive [162]. Here, different behaviours are observed for the samples milled with $TiCl_3$ or TiF_3. The better re-hydrogenation yield is achieved using TiF_3 as additive. This is the first report of such difference observed between additives having similar electronic structure, which is in agreement with the slightly higher reaction yield obtained during ball milling. The difference observed is in particular related to the formation of an unknown phase (Ca-H-Cl) in the sample containing $TiCl_3$, that prevents the formation of $Ca(BH_4)_2$. When using TiF_3, the F anion is present in CaF_2 formed during milling and does not form any other by-products. So $Ca(BH_4)_2$ is formed during heating at 350°C and 100 bar H_2. After 50 h, the reversibility yield is 67% and is limited by the slow kinetics and the formation of Ca-F compounds. From the results presented here, it is not possible to fully describe the hydrogenation process. A two steps process is assumed from the hydrogenation advancement as a function of time showing a first fast H_2 absorption step followed by a second step with lower absorption rate. However, no intermediate phase is detected in the XRD pattern of the sample hydrogenated for 50 h at 100bar and 350°C whereas the hydrogenation is not complete. In addition to the absence of by-products, the better hydrogenation properties of the sample milled with TiF_3 may be explain by a modification of the thermodynamic properties by a substitution of H by F in $Ca(BH_4)_2$ and/or by an improvement of the kinetics by Ti species (e.g. TiB_2), $CaH_{2-x}F_x$ or eventually Ti-F-Ca phases formed after decomposition of TiF_3. Further studies of the chemical and structural properties of the additives after milling and cycling are necessary to understand the catalytic mechanism. This is of great importance for the further improvement of the sorption properties of $Ca(BH_4)_2$.

5. Summary and outlook

In this work, two different types of complex hydrides are studied regarding their H_2 sorption performances. $NaAlH_4$ and $Ca(BH_4)_2$ were prepared by mechano-synthesis from their decomposition products using the reactive ball milling technique. In the first part, the reaction and catalytic mechanisms for doped **$NaAlH_4$** are studied in detail. In the second part, the synthesis and hydrogen sorption properties of $Ca(BH_4)_2$ are described.

The first part of this work is focussed on the analysis of doped $NaAlH_4$. A very simple and efficient synthesis route is established by reactive ball milling of NaH, Al and a dopant under H_2 atmosphere. In-situ monitoring of the pressure and the temperature during the process is used to bring new insight into the milling process. It is possible to observe the different parameters influencing the synthesis and obtain optimised milling conditions. An important issue for $NaAlH_4$ is the influence of the dopant that is not fully understood, even for the most studied compound $TiCl_3$. In addition, there are little studies of the doping mechanism for recently proposed dopants $ScCl_3$ and $CeCl_3$. The in-situ monitoring tool is found to be very useful for a first screening of dopants with $TiCl_3$ disclosed as the most efficient compound for the synthesis reaction, followed by $CeCl_3$ and $ScCl_3$ with lower efficiency. This result is confirmed by a further analysis of the first decomposition of as-milled $NaAlH_4$. On the other hand, on the re-hydrogenation step, a different trend is found with $CeCl_3$ being the most efficient dopant. The different dopants do not have any influence on the thermodynamic properties of $NaAlH_4$ and their role is purely kinetic. The observation that different compounds are efficient for the decomposition or the rehydrogenation processes indicates that the reaction mechanisms of these processes are quite different.

Going a step further from these direct observations, the reaction mechanism is investigated by different techniques. For the desorption process, the chemical states of the dopants showing different efficiencies are compared. From the results obtained, the main limiting step proposed for decomposition is the diffusion of hydrogen atoms or hydrogenated species through the vacancies created during milling, as already suggested by theoretical studies. In particular, the vacancies are created by the decomposition of the dopant itself and the formation of NaCl. It is therefore necessary to ensure a complete decomposition of the dopant during milling that is not possible in the case of $ScCl_3$ and $CeCl_3$ because of their higher stability. It appears also that the presence of Al-M phases (M=Ti, Sc or Ce) is important for the reaction, probably by acting as H traps similarly to the vacancies or/and as grain refiner to ensure the preservation of the nanocrystalline structure. The analysis of the re-hydrogenation mechanism is more difficult, but a crucial point is the formation of Al-M-H complexes that most likely favour the absorption of H atoms in the bulk, the formation of Ce-Al-H being particularly suitable. Besides, the formation of Ce-based compounds at the surface of the particle may also allow a more efficient splitting of the H_2 molecule at the surface, participating to the fast hydrogenation process.

In this work, new insights are given for the catalytic mechanism describing the desorption and absorption processes. This is the first report of a systematic study of both reaction processes; most of the work already published described only the desorption process. Nevertheless, additional work is necessary to complete the analysis presented here. The reaction and catalytic mechanisms occurring during re-hydrogenation are studied in less detail than the decomposition of $NaAlH_4$. More work is planned, in particular on the chemical state of the dopants or on the action of Al_4Ce, to fully understand the reactions involved. In addition, the preparation of $NaAlH_4$ co-doped with a mixture of $TiCl_3$ and $CeCl_3$ is expected to lead to enhanced kinetic properties in both absorption and desorption considering the respective efficiencies of $CeCl_3$ and $TiCl_3$ for these processes. The initial results obtained by mixing 2 mol% $TiCl_3$ and 2 mol% $CeCl_3$ show an improved performance in desorption compared to the sample milled with $CeCl_3$ only and in absorption compared to the sample milled with $TiCl_3$. Further work is in progress to optimise the dopant mixture, in particular by reducing the quantity of dopant added and to fully confirm the catalytic mechanism proposed here.

In the second part concerning the study of $Ca(BH_4)_2$, the first objective was to find a simple and fast synthesis technique. Reactive ball milling is used because it shows a high effectiveness for the synthesis of $NaAlH_4$. Nevertheless, the direct application of the knowledge obtained by analysing the synthesis of $NaAlH_4$ is not possible because of the inertness of one of the starting compounds, CaB_6, and the much slower reaction rates observed for the borohydride compounds. After 25 h milling under severe milling conditions, the yield of $Ca(BH_4)_2$ in the sample is limited (20-30%) and the combination of several analysis techniques is necessary to confirm the formation of this compound because it is poorly crystallised after milling. The analysis of the kinetic properties of the doped $Ca(BH_4)_2$ samples shows that this compound is very sensitive to the atmosphere used for measurement because different decomposition paths are found according to the atmosphere (H_2, Ar or vacuum). This is mainly related to the strong kinetic limitations that do not allow reaching the equilibrium conditions and also likely to the formation of different intermediate phases. This is in agreement with the results already reported before that show the formation of different phases according to the decomposition conditions. In addition, it appears that the decomposition route is more complicated than the two decomposition steps initially reported, with for example, an overlapping of different reactions occurring at the same temperature that are hardly distinguishable by DSC measurements. The XRD analysis of the phases present after heating the as-milled sample with TiF_3 to different temperatures (under H_2 atmosphere) shows that $Ca(BH_4)_2$ prepared by reactive ball milling follows the same decomposition route as pure $Ca(BH_4)_2$ (under these conditions), confirming the validity of the reactive ball milling technique for synthesis. The decomposition occurs in three steps. The first step leads to the formation of CaB_2H_x and CaH_2, but after the second and third steps the intermediate phase(s) and final products are not fully identified. The first two steps are not distinguishable for pure $Ca(BH_4)_2$, since they may overlap in this case. For the identification of the intermediate and final products, it will be necessary to use complementary characterisation techniques. In addition, in contrast to the results reported for $NaAlH_4$, the anion of the additive appears to play a very important role for the re-hydrogenation to $Ca(BH_4)_2$ of the decomposed sample. The re-hydrogenation is only achieved using TiF_3 but not $TiCl_3$. This is surprising considering that these two compounds have similar electronic structures. The absence of reversibility for the sample milled with $TiCl_3$ is mainly related to the formation of another phase Ca-H-Cl that hinders the formation of $Ca(BH_4)_2$. An additional effect of the F anion is also considered, possibly linked to the formation of CaF_2, but has not been verified yet. The re-hydrogenation of the sample milled with TiF_3 shows a very good yield (67% at 100 bar H_2 and 350°C after 50 h) similar to what has been reported before for similar conditions. The low reaction yield obtained during milling does not limit further cycling of $Ca(BH_4)_2$.

This work is the first report on the synthesis of $Ca(BH_4)_2$ using a room temperature method, i.e. reactive ball milling. In addition, new information about the decomposition route and the re-hydrogenation process is given, e.g. by showing more than two reaction steps for decomposition. The study of the re-hydrogenation process shows that the nature of the additive greatly influences the hydrides formed. So far, only the desorption process has been studied in detail and most of the work published concerns the crystal structure of $Ca(BH_4)_2$. A complete overview of the sorption properties of $Ca(BH_4)_2$ could not be made in this work, mainly because of the poor kinetics. The discovery of more efficient additives is then a crucial issue. In particular, understanding the beneficial role of TiF_3 is a first step towards a better understanding of the doping mechanism. Once the kinetic barrier is overcome, a more complete analysis of this system should be possible as it is done for $NaAlH_4$. This would be an important step towards a practical application of $Ca(BH_4)_2$. The determination of the thermodynamic properties by investigating the desorption and absorption steps is necessary to verify the potential application of $Ca(BH_4)_2$ as hydrogen storage material.

As a more general conclusion on solid state hydrogen storage, it appears that reaching the targets given by the DOE will be difficult in the next years. These targets have been reset to lower values in 2009 considering the quite ambitious ones given initially. The DOE targets and the future of solid state hydrogen storage are still a matter of intense debate. $NaAlH_4$ is the most advanced compound at the moment and is used as a reference system, also for the development of prototype storage tanks. It is certainly worth to start testing such compounds in tanks to demonstrate the feasibility of hydrogen for mobile applications. At present, the H_2 capacity is too low for a long driving range but the tanks developed now should prepare the technology for the use of more suitable hydrides. The first results published reveal the technological and engineering bottlenecks (e.g. upscaling the $NaAlH_4$ production, heat management, compressibility of the powder) to be mastered for the construction of complex hydride tanks [206].

Together with the tanks, new materials should be developed with higher performance than $NaAlH_4$. Different strategies are considered to improve solid state hydrogen storage materials and bring them closer to real applications, in particular based on metal borohydride compounds having high H_2 capacities. As mentioned for $Ca(BH_4)_2$, there is still a large research effort to be undertaken to improve the H_2 sorption properties of pure metal borohydrides that are far from practical applications. At present, they still suffer from very slow kinetics and too few experimental investigations of the thermodynamic properties to confirm that they can be used at moderate temperatures. For example, the first report of the enthalpy of reaction for the desorption of $Ca(BH_4)_2$ is much higher than predicted by first-principles calculations [143] but has to be confirmed by measuring the enthalpy for absorption. It is also necessary to identify the intermediate steps and compounds, which are often reported for the decomposition of metal borohydrides. In particular, the formation of $M_xB_{12}H_{12}$ compounds or amorphous boron appear to be detrimental for the re-hydrogenation process [142, 159, 207]. This is mainly related to the strong B-B bonds formed that are difficult to split for forming B-H bonds in borohydrides.

In addition to the investigation of new compounds and dopants, different concepts have been proposed in recent years that bring interesting new ideas to enhance solid state storage performance. For example, it has been shown that the thermodynamic properties of complex hydrides can be tuned by adding a metal hydride to form a so-called reactive hydride composite (RHC) [36, 43]. The addition of the metal hydride modifies the H_2 sorption reactions, in particular with the formation of borides rather than elemental boron during decomposition. The overall reaction has a lower enthalpy of reaction than the decomposition of the pure compound. Using this concept, some complex hydrides like $LiBH_4$, excluded because of their high enthalpy of reaction, can be used under much moderate conditions by adding MgH_2 [37]. Nevertheless, the decomposition occurs in two different steps with, first, the decomposition of MgH_2 and then of $LiBH_4$, but with formation of MgB_2 rather than elemental boron. The thermodynamics of these two reactions should be considered for applications because it differs in the values given for the overall decomposition reaction. However, the study of different additives leads to dramatically enhanced kinetic properties because of the obtention of a nanostructure favouring the diffusion processes and of small boride clusters favouring the nucleation processes [38].

Alternatively, following the relation that has been established between the electronegativity of the cation in the metal borohydride and the decomposition temperature, new compounds with mixed cations have been proposed, e.g. $LiAl(BH_4)_4$ [208-210]. The mixture of cations is chosen to tune the electronegativity to the value corresponding to the desired decomposition temperature. The initial results published show a decomposition temperature below 100°C for $LiAl(BH_4)_4$ or $LiZn(BH_4)_3$ [208] but unfortunately the authors have also observed a simultaneous release of diborane B_2H_6. The release of such a very toxic compound has to be eluded when using mixed cation borohydrides for solid state storage, in particular because it hinders the re-hydrogenation process as B atoms are lost. In addition, the structures of these compounds appear more complicated than initially reported with for example the formation of large anions $[Sc(BH_4)_4]^-$ counterbalanced by Na^+ in $NaSc(BH_4)_4$ [211]. Furthermore, the synthesis method used, based on a metathesis reaction between $LiBH_4$ or $NaBH_4$ and a metal chloride, is not as straightforward as initially reported [35] with the insertion of some Cl atoms in the borohydride structure in some cases [211, 212] modifying the properties of the borohydride compounds [213, 214].

Finally, another interesting approach proposed is the nanoconfinement of the complex hydrides, for example in a carbon nanoscaffold. The reduction of size to the nanometer scale appears to improve the kinetic properties (by reducing the reaction paths) but may also change the thermodynamic properties [182]. Promising results have already been obtained for $NaAlH_4$ [215, 216], $LiBH_4$ [217, 218] and $Mg(BH_4)_2$ [219, 220]. The main drawback of this confinement is the decrease of the H_2 capacity because of the weight and volume of the necessary nanoscaffolds. The decomposition temperature is dramatically decreased for undoped compounds, i.e. at 150°C in one step for $NaAlH_4$ whereas the bulk undoped $NaAlH_4$ releases H_2 between 200 and 300°C in two steps [221]. The re-absorption process is also improved by the limitation of the segregation of the different decomposition products [182, 217]. The modification of the thermodynamic properties, shown by the modification of the stability range of $NaAlH_4$ [215] or the absence of an equilibrium pressure plateau [216], opens the possibility of modifying the properties of the very stable borohydride compounds, containing a large quantity of H_2 that would preserve a reasonable H_2 capacity after confinement.

All the results presented in this work underline the difficulty of understanding the reaction mechanism for hydrogen sorption in complex hydrides. Even for the most studied one, $NaAlH_4$, some unclear mechanisms remain. The development of hydrogen storage materials for mobile application appears therefore really challenging. The more interesting compound with high H_2 capacity, namely the metal borohydrides, are particularly difficult to investigate as shown for $Ca(BH_4)_2$. In particular, great care should be taken on the measurement conditions to ensure a valid analysis. Nevertheless, there are some possible routes to improve the properties of these compounds as shown above and a continuous research effort should provide new solutions and materials for solid state hydrogen storage.

6. References

(1) Züttel, A.; Borgschulte, A.; Schlapbach, L. *Hydrogen as a future energy carrier,* Wiley-VCH: Weinheim, Germany, **2008**.

(2) Winter, C. J. *Hydrogen energy - Abundant, efficient, clean: a debate over the energy-system-of-change;* Int. J. Hydrog. Energy **2009**, *34*, S1-S52.

(3) Grove, W. R. *Unknown;* Philos. Mag. III **1839**, *14*, 127.

(4) von Helmolt, R.; Eberle, U. *Fuel cell vehicles: status 2007;* J. Power Sources **2007**, *165*, 833-843.

(5) Lefèvre, M.; Proietti, E.; Jaouen, F.; Dodelet, J. P. *Iron-based catalysts with improved oxygen reduction activity in polymer electrolyte fuel cells;* Science **2009**, *324*, 71-74.

(6) Schlapbach, L. *Hydrogen-fuelled vehicles;* Nature **2009**, *460*, 809-811.

(7) Schlapbach, L.; Züttel, A. *Hydrogen-storage materials for mobile applications;* Nature **2001**, *414*, 353-358.

(8) Eberle, U.; Felderhoff, M.; Schüth, F. *Chemical and physical solutions for hydrogen storage;* Angew. Chem. Int. Ed **2009**, *48*, 6608-6630.

(9) DOE Hydrogen and Fuel Cells, www1.eere.energy.gov/hydrogenandfuelcells/storage/index..html, **2009**

(10) Züttel, A. *Materials for hydrogen storage;* Materials Today **2003**, *6*, 24-33.

(11) Hirscher, M.; Panella, B. *Hydrogen storage in metal-organic frameworks;* Scr. Mater. **2007**, *56*, 809-812.

(12) Okada, Y.; Sasaki, E.; Watanabe, E.; Hyodo, S.; Nishijima, H. *Development of dehydrogenation catalyst for hydrogen generation in organic chemical hydride method;* Int. J. Hydrog. Energy **2006**, *31*, 1348-1356.

(13) Alapati, S. V.; Johnson, J. K.; Sholl, D. S. *Identification of destabilized metal hydrides for hydrogen storage using first principles calculations;* J. Phys. Chem. B **2006**, *110*, 8769-8776.

(14) Borgschulte, A.; Boesenberg, U.; Barkhordarian, G.; Dornheim, M.; Bormann, R. *Enhanced hydrogen sorption kinetics of magnesium by destabilized $MgH_{2-\delta}$;* Catal. Today **2007**, *120*, 262-269.

(15) Dornheim, M.; Doppiu, S.; Barkhordarian, G.; Boesenberg, U.; Klassen, T.; Gutfleisch, O.; Bormann, R. *Hydrogen storage in magnesium-based hydrides and hydrides composites;* Scr. Mater. **2007**, *56*, 841-846.

(16) Gutfleisch, O.; Schlorke-de Boer, N.; Ismail, N.; Herrich, M.; Walton, A.; Speight, J.; Harris, I. R.; Pratt, A.; Züttel, A. *Hydrogenation properties of nanocrystalline Mg- and Mg_2Ni-based compounds modified with platinum group metals (PGMs);* J. Alloys Compd **2003**, *356-357*, 598-602.

(17) Barkhordarian, G.; Klassen, T.; Bormann, R. *Kinetic investigation of the effect of milling time on the hydrogen sorption reaction of magnesium catalyzed with different Nb_2O_5 contents;* J. Alloys Compd **2006**, *407*, 249-255.

(18) Bogdanovic, B.; Schwickardi, M. *Ti-doped alkali metal aluminium hydrides as potential novel reversible hydrogen storage materials;* J. Alloys Compd **1997**, *253-254*, 1-9.

(19) Orimo, S.; Nakamori, Y.; Eliseo, J. R.; Züttel, A.; Jensen, C. M. *Complex hydrides for hydrogen storage;* Chem. Rev. **2007**, *107*, 4111-4132.

(20) Chen, P.; Xiong, Z.; Luo, J.; Lin, J.; Tan, K. L. *Interaction of hydrogen with metal nitrides and imides;* Nature **2002**, *420*, 302-304.

(21) Chen, J.; Kuriyama, N.; Xu, Q.; Takeshita, H. T.; Sakai, T. *Reversible hydrogen storage via Titanium-catalysed $LiAlH_4$ and Li_3AlH_6;* J. Phys. Chem. B **2001**, *105*, 11214-11220.

(22) Blanchard, D.; Brinks, H. W.; Hauback, B. C.; Norby, P. *Desorption of $LiAlH_4$ with Ti- and V-based additives;* Mater. Sci. Eng. B **2004**, *108*, 54-59.

(23) Palumbo, M.; Torres, F. J.; Ares, J. R.; Pisani, C.; Fernandez, J. F.; Baricco, M. *Thermodynamic and ab initio investigation of the Al-H-Mg system.;* Calphad-Comput. Coupling Ph. Diagrams Thermochem. **2007**, *31*, 457-467.

(24) Varin, R. A.; Chiu, Ch.; Czujko, T.; Wronski, Z. *Feasibility study of the direct mechano-chemical synthesis of nanostructured magnesium tetrahydroaluminate (alanate) [$Mg(AlH_4)_2$] complex hydride;* Nanotechnology **2005**, *16*, 2261-2274.

(25) Marashdeh, A.; Frankcombe, T. J. *$Ca(AlH_4)_2$, $CaAlH_5$ and CaH_2 + $LiBH_4$: calculated dehydrogenation enthalpy, including zero point energy, and the structure of the phonon spectra;* J. Chem. Phys. **2008**, *128*, 234505.

(26) Züttel, A.; Wenger, P.; Sudan, P.; Mauron, Ph.; Emmenegger, Ch. *$LiBH_4$ a new hydrogen storage material;* J. Power Sources **2003**, *118*, 1-7.

(27) Zhang, Y.; Zhang, W. S.; Fan, M. Q.; Liu, S. S.; Chu, H. L.; Zhang, Y. H.; Gao, X. Y.; Sun, L. X. *Enhanced hydrogen storage performance of $LiBH_4$-SiO_2-TiF_3 composite;* J. Phys. Chem. C **2008**, *112*, 4005-4010.

(28) Garroni, S.; Pistidda, C.; Brunelli, M.; Vaughan, G. B. M.; Surinach, S.; Baro, M. D. *Hydrogen desorption mechanism of 2 $NaBH_4$ + MgH_2 composite prepared by high energy ball milling;* Scr. Mater. **2009**, *60*, 1129-1132.

(29) Mauron, Ph.; Buchter, F.; Friedrichs, O.; Remhof, A.; Bielmann, M.; Zwicky, C. N.; Züttel, A. *Stability and reversibility of $LiBH_4$;* J. Phys. Chem. B **2008**, *112*, 906-910.

(30) Grochala, W.; Edwards, P. P. *Thermal decomposition of the non-interstitial hydrides for the storage and production of hydrogen;* Chem. Rev. **2004**, *104*, 1283-1315.

(31) Miwa, K.; Aoki, M.; Noritake, T.; Ohba, N.; Nakamori, Y.; Towata, S.; Züttel, A.; Orimo, S. *Thermodynamical stability of calcium borohydride $Ca(BH_4)_2$;* Phys. Rev. B **2006**, *74*, 155122.

(32) Chlopek, K.; Frommen, C.; Léon, A.; Zabara, O.; Fichtner, M. *Synthesis and properties of magnesium tetrahydroborate, $Mg(BH_4)_2$;* J. Mater. Chem. **2007**, *17*, 3496-3503.

(33) Li, H. W.; Kikuchi, K.; Nakamori, Y.; Miwa, K.; Towata, S.; Orimo, S. *Effects of ball milling and additives on dehydriding behaviors of well-crystallized $Mg(BH_4)_2$;* Scr. Mater. **2007**, *57*, 679-682.

(34) Soloveichik, G. L.; Gao, Y.; Rijssenbeek, J.; Andrus, M.; Kniajanski, S.; Bowman, R. C.; Hwang, S. J.; Zhao, J. C. *Magnesium borohydride as a hydrogen storage material: properties and dehydrogenation pathway of unsolvated $Mg(BH_4)_2$.;* Int. J. Hydrog. Energy **2009**, *34*, 916-928.

(35) Nakamori, Y.; Li, H. W.; Kikuchi, K.; Aoki, M.; Miwa, K.; Towata, S.; Orimo, S. *Thermodynamical stabilities of metal-borohydrides;* J. Alloys Compd **2007**, *446-447*, 296-300.

(36) Vajo, J. J.; Skeith, S. L.; Mertens, F. *Reversible storage of hydrogen in destabilized $LiBH_4$;* J. Phys. Chem. B Lett. **2005**, *109*, 3719-3722.

(37) Bösenberg, U.; Doppiu, S.; Mosegaard, L.; Barkhordarian, G.; Eigen, N.; Borgschulte, A.; Jensen, T. R.; Cerenius, Y.; Gutfleisch, O.; Klassen, T.; Dornheim, M.; Bormann, R. *Hydrogen sorption properties of MgH_2-$LiBH_4$ composites;* Acta Mater. **2007**, *55*, 3951-3958.

(38) Bösenberg, U.; Kim, J. W.; Gosslar, D.; Eigen, N.; Jensen, T. R.; Bellosta von Colbe, J. M.; Zhou, Y.; Dahms, M.; Kim, D. H.; Günther, R.; Cho, Y. W.; Oh, K. H.; Klassen, T.; Bormann, R.; Dornheim, M. *Role of additives in $LiBH_4$-MgH_2 reactive hydride composites for sorption kinetics;* Acta Mater. **2010**, *58*, 3381-3389.

(39) Deprez, E.; Munoz-Marquez, M. A.; Roldan, M. A.; Prestipino, C.; Javier Palomares, F.; Bonatto Minella, C.; Bösenberg, U.; Dornheim, M.; Bormann, R.; Fernandez, A. *Oxidation state and local structure of Ti-based additives in the reactive hydride composite $2LiBH_4$ + MgH_2;* J. Phys. Chem. C **2010**, *114*, 3309-3317.

(40) Dornheim, M.; Eigen, N.; Barkhordarian, G.; Klassen, T.; Bormann, R. *Tailoring hydrogen storage materials towards application;* Adv. Eng. Mater. **2006**, *8*, 377-385.

(41) Ozolins, V.; Majzoub, E.; Wolverton, C. *First-principles prediction of thermodynamically reversible hydrogen storage reactions in the Li-Mg-Ca-B-H system;* J. Am. Chem. Soc. **2009**, *131*, 230-237.

(42) Siegel, D. J.; Wolverton, C.; Ozolins, V. *Thermodynamic guidelines for the prediction of hydrogen storage reactions and their application to destabilized hydride mixtures;* Phys. Rev. B **2007**, *76*, 134102.

(43) Barkhordarian, G.; Jensen, T. R.; Doppiu, S.; Bösenberg, U.; Borgschulte, A.; Gremaud, R.; Cerenius, Y.; Dornheim, M.; Klassen, T.; Bormann, R. *Formation of $Ca(BH_4)_2$ from hydrogenation of CaH_2 + MgB_2;* J. Phys. Chem. C **2008**, *112*, 2743-2749.

(44) Kim, Y.; Reed, D.; Lee, Y. S.; Shim, J. H.; Han, H. N.; Book, D.; Cho, Y. W. *Hydrogenation reaction of CaH_2 - CaB_6 - Mg mixture;* J. Alloys Compd **2010**, *492*, 597-600.

(45) Shim, J. H.; Lim, J. H.; Rather, S.; Lee, Y. S.; Reed, D.; Kim, Y.; Book, D.; Cho, Y. W. *Effect of hydrogen back pressure on dehydrogenation behavior of $LiBH_4$-based reactive hydride composite;* J. Phys. Chem. Lett. **2010**, *1*, 59-63.

(46) Ross, D. J.; Halls, M. D.; Nazri, A. G.; Aroca, R. F. *Raman scattering of complex sodium aluminium hydride for hydrogen storage;* Chem. Phys. Lett. **2004**, *388*, 430-435.

(47) Opalka, S. M.; Anton, D. L. *First principles study of sodium-aluminium-hydrogen phases;* J. Alloys Compd **2003**, *356-357*, 486-489.

(48) Bastide, J. P.; Bonnetot, B.; Létoffé, J. M.; Claudy, P. *Polymorphisme de l'hexahydroaluminate trisodique Na_3AlH_6;* Met. Res. Bull. **1981**, *16*, 91-96.

(49) Claudy, P.; Bonnetot, B.; Chahine, G.; Letoffe, J. M. *Etude du comportement thermique du tertahydroaluminate de sodium $NaAlH_4$ et de l'hexahydroaluminate de sodium Na_3AlH_6 de 298 à 600K.;* Thermochim. Acta **1980**, *38*, 75-88.

(50) Bogdanovic, B.; Brand, R. A.; Marjanovic, A.; Schwickardi, M.; Tölle, J. *Metal-doped sodium aluminium hydrides as potential new hydrogen storage materials;* J. Alloys Compd **2000**, *302*, 36-58.

(51) Gross, K. J.; Thomas, G. J.; Jensen, C. M. *Catalysed alanates for hydrogen storage;* J. Alloys Compd **2002**, *330-332*, 683-690.

(52) Lee, B. M.; Jang, J. W.; Shim, J. H.; Cho, Y. W.; Lee, B. J. *Thermodynamic assessment of the NaH <-> Na_3AlH_6 <-> $NaAlH_4$ hydride system;* J. Alloys Compd **2006**, *424*, 370-375.

(53) Peles, A.; Alford, A.; Ma, Z.; Yang, L.; Chou, M. Y. *First-principles study of $NaAlH_4$ and Na_3AlH_6 complex hydrides;* Phys. Rev. B **2004**, *70*, 165105.

(54) Chung, S. C.; Morioka, H. *Thermochemistry and crystal structures of lithium, sodium and potassium alanates as determined by ab initio simulations;* J. Alloys Compd **2004**, *372*, 92-96.

(55) Alapati, S. V.; Johnson, J. K.; Sholl, D. S. *Predicting reaction equilibria for destabilized metal hydride decomposition reactions for reversible hydrogen storage.;* J. Phys. Chem. C **2007**, *111*, 1584-1591.

(56) Zaluski, L.; Zaluska, A.; Ström-Olsen, J. O. *Hydrogenation properties of complex alkali metal hydrides fabricated by mechano-chemical synthesis;* J. Alloys Compd **1999**, *290*, 71-78.

(57) Dilts, J. A.; Ashby, E. C. *A study of the thermal decomposition of complex metal hydrides;* Inorg. Chem. **1972**, *11*, 1230-1236.

(58) Gross, K. J.; Sandrock, G.; Thomas, G. J. *Dynamic in situ X-ray diffraction of catalyzed alanates;* J. Alloys Compd **2002**, *330-332*, 691-695.

(59) Gross, K. J.; Guthrie, S.; Takara, S.; Thomas, G. *In-situ X-ray diffraction study of the decomposition of $NaAlH_4$;* J. Alloys Compd **2000**, *297*, 270-281.

(60) Kim, J. W.; Shim, J. H.; Kim, S. C.; Remhof, A.; Borgschulte, A.; Friedrichs, O.; Gremaud, R.; Pendolino, F.; Züttel, A.; Cho, Y. W.; Oh, K. H. *Catalytic effect of TiN nanopowder on hydrogen desorption properties of $NaAlH_4$ and its stability in $NaAlH_4$;* J. Power Sources **2009**, *192*, 582-587.

(61) Lee, G. J.; Shim, J. H.; Cho, Y. W.; Lee, K. S. *Reversible hydrogen storage in $NaAlH_4$ catalyzed with lanthanide oxides;* Int. J. Hydrog. Energy **2007**, *32*, 1911-1915.

(62) Schüth, F.; Bogdanovic, B.; Felderhoff, M. *Light metal hydrides and complex hydrides for hydrogen storage;* Chem. Commun. **2004**, 2249-2258.

(63) Sandrock, G.; Gross, K. J.; Thomas, G. *Effect of Ti-catalyst content on the reversible hydrogen storage properties of the sodium alanates;* J. Alloys Compd **2002**, *339*, 299-308.

(64) Kircher, O.; Fichtner, M. *Kinetic studies of the decomposition of $NaAlH_4$ doped with a Ti-based catalyst.;* J. Alloys Compd **2005**, *404-406*, 339-342.

(65) Anton, D. L. *Hydrogen desorption kinetics in transition metal modified $NaAlH_4$;* J. Alloys Compd **2003**, *356-357*, 400-404.

(66) Mueller, W. M.; Blackledge, J. P.; Libowitz, G. G. *Metal Hydrides,* Academic Press: New York, **1968**.

(67) Srinivasan, S. S.; Brinks, H. W.; Hauback, B. C.; Sun, D.; Jensen, C. M. *Long term cycling behavior of titanium doped $NaAlH_4$ prepared through solvent mediated milling of NaH and Al with titanium precursors;* J. Alloys Compd **2004**, *377*, 283-289.

(68) Jensen, C. M.; Zidan, R.; Mariels, N.; Hee, A.; Hagen, C. *Advanced titanium doping of sodium aluminium hydride: segue to a practical hydrogen storage materials?;* Int. J. Hydrog. Energy **1999**, *24*, 461-465.

(69) Zidan, R. A.; Takara, S.; Hee, A.; Jensen, C. M. *Hydrogen cycling behavior of zirconium and titanium-zirconium-doped sodium aluminum hydride;* J. Alloys Compd **1999**, *285*, 119-122.

(70) Onkawa, M.; Zhang, S.; Takeshita, H. T.; Kuriyama, N.; Kiyobayashi, T. *Dehydrogenation kinetics of Ti-doped $NaAlH_4$ - Influence of Ti precursors and preparation methods.;* Int. J. Hydrog. Energy **2008**, *33*, 718-721.

(71) Meisner, G. P.; Tibbetts, G. G.; Pinkerton, F. E.; Olk, C. H.; Balogh, M. P. *Enhancing low pressure hydrogen storage in sodium alanates;* J. Alloys Compd **2002**, *337*, 254-263.

(72) Zaluska, A.; Zaluski, L.; Ström-Olsen, J. O. *Sodium alanates for reversible hydrogen storage;* J. Alloys Compd **2000**, *298*, 125-134.

(73) Resan, M.; Hampton, M. D.; Lomness, J. K.; Slaterry, D. K. *Effect of Ti_xAl_y catalysts on hydrogen storage properties of $LiAlH_4$ and $NaAlH_4$;* Int. J. Hydrog. Energy **2005**, *30*, 1417-1421.

(74) Gross, K. J.; Majzoub, E.; Spangler, S. W. *The effects of titanium precursors on hydriding properties of alanates;* J. Alloys Compd **2003**, *356-357*, 423-428.

(75) Eigen, N.; Kunowsky, M.; Klassen, T.; Bormann, R. *Synthesis of $NaAlH_4$-based hydrogen storage material using milling under low pressure hydrogen atmosphere;* J. Alloys Compd **2007**, *430*, 350-355.

(76) Bellosta von Colbe, J. M.; Felderhoff, M.; Bogdanovic, B.; Schüth, F.; Weidenthaler, C. *One-step direct synthesis of a Ti-doped sodium alanate hydrogen storage material;* Chem. Commun. **2005**, *37*, 4732-4734.

(77) Bogdanovic, B.; Felderhoff, M.; Pommerin, A.; Schüth, F.; Spielkamp, N.; Stark, A. *Cycling properties of Sc- and Ce-doped $NaAlH_4$ hydrogen storage materials prepared by the one-step direct synthesis method;* J. Alloys Compd **2009**, *471*, 383-386.

(78) Léon, A.; Kircher, O.; Rösner, H.; Décamps, B.; Leroy, E.; Fichtner, M.; Percheron-Guégan, A. *SEM and TEM characterization of sodium alanate doped with $TiCl_3$ or small Ti clusters (Ti_{13}-6THF);* J. Alloys Compd **2006**, *414*, 190-203.

(79) Thomas, G. J.; Gross, K. J.; Yang, N. Y. C.; Jensen, C. M. *Microstructural characterization of catalyzed $NaAlH_4$;* J. Alloys Compd **2002**, *330-332*, 702-707.

(80) Sandrock, G.; Gross, K. J.; Thomas, G.; Jensen, C. M.; Meeker, D.; Takara, S. *Engineering considerations in the use of catalyzed sodium alanates for hydrogen storage;* J. Alloys Compd **2002**, *330-332*, 696-701.

(81) Wang, P.; Kang, X. D.; Cheng, H. M. *Improved Hydrogen storage of TiF_3-doped $NaAlH_4$;* Chem. Phys. Chem. **2005**, *6*, 2488-2491.

(82) Kang, X. D.; Wang, P.; Cheng, H. M. *Advantage of TiF_3 over $TiCl_3$ as a dopant precursor to improve the thermodynamic property of Na_3AlH_6;* Scr. Mater. **2007**, *56*, 361-364.

(83) Yin, L. C.; Wang, P.; Kang, X. D.; Sun, C. H.; Cheng, H. M. *Functional anion concept: effect of fluorine anion on hydrogen storage of sodium alanate;* Phys. Chem. Chem. Phys. **2007**, *9*, 1499-1502.

(84) Wang, P.; Jensen, C. M. *Method for preparing Ti-doped $NaAlH_4$ using Ti powder: observation of an unusual reversible dehydrogenation behavior;* J. Alloys Compd **2004**, *379*, 99-102.

(85) Xiao, X.; Chen, L.; Wang, X.; Wang, Q.; Chen, C. *The hydrogen storage properties and microstructure of Ti-doped sodium aluminum hydride prepared by ball-milling;* Int. J. Hydrog. Energy **2007**, *32*, 2475-2479.

(86) Léon, A.; Kircher, O.; Fichtner, M.; Rothe, J.; Schild, D. *Evolution of the local structure around Ti atoms in $NaAlH_4$ doped with $TiCl_3$ or Ti_{13}-6THF by ball milling using X-ray Absorption and X-ray photoelectron spectroscopy;* J. Phys. Chem. B **2006**, *110*, 1192-1200.

(87) Fichtner, M.; Fuhr, O.; Kircher, O.; Rothe, J. *Small Ti clusters for catalysis of hydrogen exchange in $NaAlH_4$;* Nanotechnology **2003**, *14*, 778-785.

(88) Wang, T.; Wang, J.; Ebner, A. D.; Ritter, J. A. *Reversible hydrogen storage properties of $NaAlH_4$ catalyzed with scandium;* J. Alloys Compd **2008**, *450*, 293-300.

(89) Bogdanovic, B.; Felderhoff, M.; Pommerin, A.; Schüth, F.; Spielkamp, N. *Advanced Hydrogen-storage materials based on Sc-, Ce-, and Pr-doped $NaAlH_4$;* Adv. Mater. **2006**, *18*, 1198-1201.

(90) Liu, J.; Ge, Q. *A first-principles study of Sc-doped $NaAlH_4$ for reversible hydrogen storage;* J. Alloys Compd **2007**, *446-447*, 267-270.

(91) Sun, T.; Zhou, B.; Wang, H.; Zhu, M. *The effect of doping rare-earth chloride dopant on the dehydrogenation properties of $NaAlH_4$ and its catalytic mechanism;* Int. J. Hydrog. Energy **2008**, *33*, 2260-2267.

(92) Liu, J.; Han, Y.; Ge, Q. *Effect of doped transition metal on reversible hydrogen release/uptake from $NaAlH_4$;* Chem. Eur. J. **2009**, *15*, 1685-1695.

(93) Blomqvist, A.; Araujo, C. M.; Jena, P.; Ahuja, R. *Dehydrogenation from 3d-transition-metal-doped $NaAlH_4$: prediction of catalysts;* Appl. Phys. Lett. **2007**, *90*, 141904.

(94) Bowman, R. C.; Reiter, J. W.; Hwang, S. J.; Kim, C.; Kabbour, H. Characterization of complex metal hydrides by high resolution solid state NMR spectroscopy. In *Materials Issue in Hydrogen Economy. Proceedings of the Internation Symposium, Nov. 12-15, 2007*, Jena, P., Kandalam, A., Sun, Q., Eds.; World Scientific Publishing: Singapore, 2009; pp 192-202.

(95) Léon, A.; Schild, D.; Fichtner, M. *Chemical state of Ti in sodium alanate doped with $TiCl_3$ using X-ray photoelectron spectroscopy;* J. Alloys Compd **2005**, *404-406*, 766-770.

(96) Majer, G.; Stanik, E.; Valiente Banuet, L. E.; Grinberg, F.; Kircher, O.; Fichtner, M. *Effects of catalysts on the dehydriding of alanates monitored by proton NMR;* J. Alloys Compd **2005**, *404-406*, 738-742.

(97) Fichtner, M.; Canton, P.; Kircher, O.; Léon, A. *Nanocrystalline alanates - Phase transformations, and catalysts;* J. Alloys Compd **2005**, *404-406*, 732-737.

(98) Graetz, J.; Reilly, J. J.; Johnson, J.; Ignatov, A.; Tyson, T. A. *X-ray absorption study of Ti-activated sodium aluminum hydride;* Appl. Phys. Lett. **2004**, *85*, 500-502.

(99) Léon, A.; Kircher, O.; Rothe, J.; Fichtner, M. *Chemical state and local structure around Titanium atoms in $NaAlH_4$ doped with $TiCl_3$ using X-ray absorption spectroscopy;* J. Phys. Chem. B **2004**, *108*, 16372-16376.

(100) Felderhoff, M.; Klementiev, K.; Grünert, W.; Spliethoff, B.; Tesche, B.; Bellosta von Colbe, J. M.; Bogdanovic, B.; Härtel, M.; Pommerin, A.; Schüth, F.; Weidenthaler, C. *Combined TEM-EDX and XAFS studies of Ti-doped sodium alanate;* Phys. Chem. Chem. Phys. **2004**, *6*, 4369-4374.

(101) Gomes, S.; Renaudin, G.; Hagemann, H.; Yvon, K.; Sulic, M. P.; Jensen, C. M. *Effects of milling, doping and cycling of $NaAlH_4$ studied by vibrational spectroscopy and X-ray diffraction;* J. Alloys Compd **2005**, *390*, 305-313.

(102) Majzoub, E.; McCarty, K. F.; Ozolins, V. *Lattice dynamics of $NaAlH_4$ from high temperature single-crystal Raman scattering and ab initio calculations: Evidence of highly stable AlH_4^- anions.;* Phys. Rev. B **2005**, *71*, 024118.

(103) Bogdanovic, B.; Eberle, U.; Felderhoff, M.; Schüth, F. *Complex aluminium hydrides;* Scr. Mater. **2007**, *56*, 813-816.

(104) Majzoub, E.; Herberg, J. L.; Stumpf, R.; Spangler, S. W.; Maxwell, R. S. *XRD and NMR investigation of Ti-compound formation in solution-doping of sodium aluminium hydrides: solubility of Ti in $NaAlH_4$ crystals grown in THF;* J. Alloys Compd **2005**, *394*, 265-270.

(105) Brinks, H. W.; Jensen, C. M.; Srinivasan, S. S.; Hauback, B. C.; Blanchard, D.; Murphy, K. *Synchrotron X-ray and neutron diffraction studies of $NaAlH_4$ containing Ti additives;* J. Alloys Compd **2004**, *376*, 215-221.

(106) Haiduc, A. G.; Stil, H. A.; Schwarz, M. A.; Paulus, P.; Geerlings, J. J. C. *On the fate of the Ti catalyst during hydrogen cycling of sodium alanate;* J. Alloys Compd **2005**, *393*, 252-263.

(107) Bogdanovic, B.; Felderhoff, M.; Germann, M.; Härtel, M.; Pommerin, A.; Schüth, F.; Weidenthaler, C.; Zibrowius, B. *Investigation of hydrogen discharging and recharging processes of Ti-doped $NaAlH_4$ by X-ray diffraction analysis (XRD) and solid-state NMR spectroscopy;* J. Alloys Compd **2003**, *350*, 246-255.

(108) Balema, V. P.; Balema, L. *Missing pieces of the puzzle or about some unresolved issues in solid state chemistry of alkali metal aluminohydrides;* Phys. Chem. Chem. Phys. **2005**, *7*, 1310-1314.

(109) Chaudhuri, S.; Graetz, J.; Ignatov, A.; Reilly, J. J.; Muckerman, J. T. *Understanding the role of Ti in reversible hydrogen storage as sodium alanate: a combined experimental and density functional theoritical approach;* J. Am. Chem. Soc. **2006**, *128*, 11404-11415.

(110) Dathar, G. K. P.; Mainardi, D. S. *Thermodynamic profiles of Ti-doped sodium alanates;* J. Phys. Chem. C **2009**, *113*, 15051-15057.

(111) Fang, F.; Zheng, S.; Chen, G.; Sang, G.; He, B.; Wei, S.; Sun, D. *Formation of Na_3AlH_6 from NaH/Al mixture and Ti-containing catalyst.;* Acta Mater. **2009**, *57*, 1959-1965.

(112) Dobbins, T. A.; Abrecht, M.; Uprety, Y.; Moore, K. *An X-ray photoemission electron microscopy study of the formation of Ti-Al phases in 4mol% $TiCl_3$ catalyzed $NaAlH_4$ during high energy ball milling;* Nanotechnology **2009**, *20*, 204014.

(113) Graham, D. D.; Culnane, L. F.; Sulic, M. P.; Jensen, C. M.; Robertson, I. M. *Ti EELS standards for identification of catalytic species in $NaAlH_4$ hydrogen storage materials;* J. Alloys Compd **2007**, *446-447*, 255-259.

(114) Bellosta von Colbe, J. M.; Schmidt, W.; Felderhoff, M.; Bogdanovic, B.; Schüth, F. *Hydrogen-isotope scrambling on doped sodium alanate;* Angew. Chem. Int. Ed **2006**, *45*, 3663-3665.

(115) Chen, J. C.; Juanes-Marcos, J. C.; Al-Halabi, A.; Olsen, R. A.; Kroes, G. J. *A theoritical study of H_2 reacting on Ti/Al(100) surfaces;* J. Phys. Chem. C **2009**, *113*, 11027-11034.

(116) Walters, R. T.; Scogin, J. H. *A reversible hydrogen storage mechanism for sodium alanate: the role of alanes and the catalytic effect of the dopant;* J. Alloys Compd **2004**, *379*, 135-142.

(117) Wiberg, E.; Uson, R. *Zur Kenntnis eines Titan-aluminium-wasserstoffs $Ti(AlH_4)_4$;* J. Naturforschg. **1951**, *6b*, 392-393.

(118) Fu, Q. J.; Ramirez-Cuesta, A. J.; Tsang, S. C. *Molecular aluminum hydrides identified by Inelastic neutron scattering during H_2 regeneration of catalyst-doped $NaAlH_4$;* J. Phys. Chem. B **2006**, *110*, 711-715.

(119) Peles, A.; Van de Walle, C. G. *Hydrogen-related defects in sodium alanate;* J. Alloys Compd **2007**, *446-447*, 459-461.

(120) Borgschulte, A.; Züttel, A.; Hug, P.; Barkhordarian, G.; Eigen, N.; Dornheim, M.; Bormann, R.; Ramirez-Cuesta, A. J. *Hydrogen-deuterium exchange experiments to probe the decomposition reaction of sodium alanate.;* Phys. Chem. Chem. Phys. **2008**, *10*, 4045-4055.

(121) Singh, S.; Eijt, S. W. H. *Hydrogen vacancies facilitate hydrogen transport kinetics in sodium hydride nanocrystallites;* Phys. Rev. B **2008**, *78*, 224110.

(122) Lodziana, Z.; Züttel, A. *Ti cation in sodium alanate;* J. Alloys Compd **2009**, *471*, L29-L31.

(123) Vegge, T. *Equilibrium structure and Ti-catalyzed H_2 desorption in $NaAlH_4$ nanoparticles from density functional theory;* Phys. Chem. Chem. Phys. **2006**, *8*, 4853-4861.

(124) Shi, Q.; Voss, J.; Jacobsen, H. S.; Lefmann, K.; Zamponi, M.; Vegge, T. *Point defect dynamics in sodium aluminum hydrides - combined quasielastic neutron scattering and density functionnal theory study;* J. Alloys Compd **2007**, *446-447*, 469-473.

(125) Li, S.; Jena, P.; Ahuja, R. *Effect of Ti and metal vacancies on the eletronic structure, stability, and dehydrogenation of Na_3AlH_6: supercell band structure formalism and gradient-corrected density-functionnal theory;* Phys. Rev. B **2006**, *73*, 214107.

(126) Lodziana, Z.; Züttel, A.; Zielinski, P. *Titanium and native defects in $LiBH_4$ and $NaAlH_4$;* J. Phys. Condens. Matter **2008**, *20*, 465210.

(127) Palumbo, O.; Paolone, A.; Cantelli, R.; Jensen, C. M.; Sulic, M. P. *Fast H-vacancy dynamics during alanate decomposition by anelastic spectroscopy. Proposition of a model for Ti-enhanced hydrogen transport;* J. Phys. Chem. B **2006**, *110*, 9105-9111.

(128) Palumbo, O.; Cantelli, R.; Paolone, A.; Jensen, C. M.; Srinivasan, S. S. *Point defect dynamics and evolution of chemical reactions in alanates by anelastic spectroscopy;* J. Phys. Chem. B **2005**, *109*, 1168-1173.

(129) Palumbo, O.; Paolone, A.; Cantelli, R.; Jensen, C. M.; Ayabe, R. *Monitoring of chemical reactions and point defect dynamics in sodium alanates;* Mater. Sci. Eng. A **2006**, *442*, 75-78.

(130) Cantelli, R.; Palumbo, O.; Paolone, A.; Jensen, C. M.; Kuba, M. T.; Ayabe, R. *Dynamics of defects in alanates;* J. Alloys Compd **2007**, *446-447*, 260-263.

(131) Singh, S.; Eijt, S. W. H.; Huot, J.; Kockelmann, W. A.; Wagemaker, M.; Mulder, F. M. *The $TiCl_3$ catalyst in $NaAlH_4$ for hydrogen storage induces grain refinement and impacts on hydrogen vacancy formation;* Acta Mater. **2007**, *55*, 5549-5557.

(132) Gunaydin, H.; Houk, K. N.; Ozolins, V. *Vacancy-mediated dehydrogenation of sodium alanate;* Proc. Natl. Acad. Sci. U. S. A. **2008**, *105*, 3673-3677.

(133) Wan, C.; Ju, X.; Qi, Y.; Zhang, Y.; Wang, S.; Liu, X.; Jiang, L. *Synchrotron XRD and XANES studies of cerium-doped $NaAlH_4$: elucidation of doping induced struture changes and electronic state;* J. Alloys Compd **2009**, *481*, 60-64.

(134) Léon, A.; Rothe, J.; Chlopek, K.; Zabara, O.; Fichtner, M. *Fluorescence XAFS study of $NaAlH_4$ doped with Ce-based precursor.;* Phys. Chem. Chem. Phys. **2009**, *11*, 8829-8834.

(135) Fan, X.; Xiao, X.; Chen, L.; Yu, K.; Wu, Z.; Li, S.; Wang, Q. *Active species of $CeAl_4$ in the $CeCl_3$-doped sodium aluminium hydride and its enhancement on reversible hydrogen storage performance;* Chem. Commun. **2009**, *44*, 6857-6859.

(136) Züttel, A.; Borgschulte, A.; Orimo, S. *Tetrahydroborates as new hydrogen storage materials;* Scr. Mater. **2007**, *56*, 823-828.

(137) Orimo, S.; Nakamori, Y.; Kitahara, G.; Miwa, K.; Ohba, N.; Towata, S.; Züttel, A. *Dehydriding and rehydriding reactions of $LiBH_4$;* J. Alloys Compd **2005**, *404-406*, 427-430.

(138) Mosegaard, L.; Moller, B.; Jorgensen, J. E.; Boesenberg, U.; Dornheim, M.; Hanson, J. C.; Cerenius, Y.; Walker, G. S.; Jakobsen, H. J.; Besenbacher, F.; Jensen, T. R. *Intermediate phases observed during decomposition of* $LiBH_4$*;* J. Alloys Compd **2007**, *446-447*, 301-305.

(139) Orimo, S.; Nakamori, Y.; Ohba, N.; Miwa, K.; Aoki, M.; Towata, S.; Züttel, A. *Experimental studies on intermediate compound of* $LiBH_4$*;* Appl. Phys. Lett. **2006**, *89*, 021920.

(140) Ohba, N.; Miwa, K.; Aoki, M.; Noritake, T.; Towata, S. *First-principles study on the stability of intermediate compounds of* $LiBH_4$*;* Phys. Rev. B **2006**, *74*, 075110.

(141) Hwang, S. J.; Bowman, R. C.; Reiter, J. W.; Rijssenbeek, J.; Soloveichik, G. L.; Zhao, J. C.; Kabbour, H.; Ahn, C. C. *NMR confirmation for formation of* $[B_{12}H_{12}]^{2-}$ *complexes during hydrogen desorption from metal borohydride;* J. Phys. Chem. C Lett. **2008**, *112*, 3164-3169.

(142) Friedrichs, O.; Buchter, F.; Borgschulte, A.; Remhof, A.; Zwicky, C. N.; Mauron, Ph.; Bielmann, M.; Züttel, A. *Direct synthesis of* $LiBH_4$ *and* $LiBD_4$ *from the elements;* Acta Mater. **2008**, *56*, 949-954.

(143) Mao, J.; Guo, Z.; Poh, C. K.; Ranjbar, A.; Guo, Y.; Yu, X.; Liu, H. *Study on the dehydrogenation kinetics and thermodynamics of* $Ca(BH_4)_2$*;* J. Alloys Compd **2010**, *500*, 200-205.

(144) Vajeeston, P.; Ravindran, P.; Fjellvag, H. *A new series of high hydrogen content hydrogen-storage materials - a theoritical prediction;* J. Alloys Compd **2007**, *446-447*, 44-47.

(145) Noritake, T.; Aoki, M.; Matsumoto, M.; Miwa, K.; Towata, S.; Li, H. W.; Orimo, S. *Crystal structure and charge density analysis of* $Ca(BH_4)_2$*;* J. Alloys Compd **2009**, *491*, 57-62.

(146) Riktor, M. D.; Sorby, M. H.; Chlopek, K.; Fichtner, M.; Buchter, F.; Züttel, A.; Hauback, B. C. *In situ synchrotron diffraction studies of phase transitions and thermal decomposition of* $Mg(BH_4)_2$ *and* $Ca(BH_4)_2$*;* J. Mater. Chem. **2007**, *17*, 4939-4942.

(147) Fichtner, M.; Chlopek, K.; Longhini, M.; Hagemann, H. *Vibrational spectra of* $Ca(BH_4)_2$*;* J. Phys. Chem. C **2008**, *112*, 11575-11579.

(148) Buchter, F.; Lodziana, Z.; Remhof, A.; Friedrichs, O.; Borgschulte, A.; Mauron, Ph.; Züttel, A.; Sheptyakov, D.; Barkhordarian, G.; Bormann, R.; Chlopek, K.; Fichtner, M.; Sorby, M. H.; Riktor, M. D.; Hauback, B. C.; Orimo, S. *Structure of* $Ca(BD_4)_2$ *β-phase from combined neutron and synchrotron X-ray powder diffraction data and density functional calculations;* J. Phys. Chem. B **2008**, *112*, 8042-8048.

(149) Majzoub, E.; Rönnebro, E. *Crystal structures of calcium borohydride: theory and experiment;* J. Phys. Chem. C **2009**, *113*, 3352-3358.

(150) Buchter, F.; Lodziana, Z.; Remhof, A.; Friedrichs, O.; Borgschulte, A.; Mauron, Ph.; Züttel, A.; Sheptyakov, D.; Palatinus, L.; Chlopek, K.; Fichtner, M.; Barkhordarian, G.; Bormann, R.; Hauback, B. C. *Structure of the orthorombic γ-phase and phase transitions of $Ca(BD_4)_2$;* J. Phys. Chem. C **2009**, *113*, 17223-17230.

(151) Filinchuk, Y.; Rönnebro, E.; Chandra, D. *Crystal structures and phase transformations in $Ca(BH_4)_2$;* Acta Mater. **2009**, *57*, 732-738.

(152) George, L.; Drozd, V.; Saxena, S. K.; Gil Bardaji, E.; Fichtner, M. *High-pressure investigation on calcium borohydride;* J. Phys. Chem. C **2009**, *113*, 15087-15090.

(153) Nakamori, Y.; Li, H. W.; Matsuo, M.; Miwa, K.; Towata, S.; Orimo, S. *Development of metal borohydrides for hydrogen storage;* J. Phys. Chem. Solids **2008**, *69*, 2292-2296.

(154) Köster, R. *Neue Herstellungsmethoden für Metallborhydride;* Angew. Chem. **1957**, *3*, 94.

(155) Rönnebro, E.; Majzoub, E. *Calcium borohydride for hydrogen storage: catalysis and reversibility;* J. Phys. Chem. B Lett. **2007**, *111*, 12045-12047.

(156) Kim, J. H.; Jin, S. A.; Shim, J. H.; Cho, Y. W. *Thermal decomposition behavior of calcium borohydride $Ca(BH_4)_2$;* J. Alloys Compd **2008**, *461*, L20-L22.

(157) Aoki, M.; Miwa, K.; Noritake, T.; Ohba, N.; Matsumoto, M.; Li, H. W.; Nakamori, Y.; Towata, S.; Orimo, S. *Structural and dehydriding properties of $Ca(BH_4)_2$;* Appl. Phys. A **2008**, *92*, 601-605.

(158) Kim, Y.; Reed, D.; Lee, Y. S.; Lee, J. Y.; Shim, J. H.; Book, D.; Cho, Y. W. *Identification of the dehydrogenated product of $Ca(BH_4)_2$;* J. Phys. Chem. C **2009**, *113*, 5865-5871.

(159) Wang, L.; Graham, D. D.; Robertson, I. M.; Johnson, D. D. *On the reversibility of hydrogen storage reactions in $Ca(BH_4)_2$: characterization via experiment and theory;* J. Phys. Chem. C **2009**, *113*, 20088-20096.

(160) Riktor, M. D.; Sorby, M. H.; Chlopek, K.; Fichtner, M.; Hauback, B. C. *The identification of a hitherto unknown intermediate phase CaB_2H_x from decomposition of $Ca(BH_4)_2$;* J. Mater. Chem. **2009**, *19*, 2754-2759.

(161) Kim, J. H.; Shim, J. H.; Cho, Y. W. *On the reversibility of hydrogen storage in Ti- and Nb-catalyzed $Ca(BH_4)_2$;* J. Power Sources **2008**, *181*, 140-143.

(162) Kim, J. H.; Jin, S. A.; Shim, J. H.; Cho, Y. W. *Reversible hydrogen storage in calcium borohydride $Ca(BH_4)_2$;* Scr. Mater. **2008**, *58*, 481-483.

(163) Lee, J. Y.; Ravnsbaek, D.; Lee, Y. S.; Kim, Y.; Cerenius, Y.; Shim, J. H.; Jensen, T. R.; Hur, N. H.; Cho, Y. W. *Decomposition reactions and reversibility of the $LiBH_4$-$Ca(BH_4)_2$ composite.;* J. Phys. Chem. C **2009**, *113*, 15080-15086.

(164) Doppiu, S.; Schultz, L.; Gutfleisch, O. *In situ pressure and temperature monitoring during the conversion of Mg into MgH_2 by high-pressure reactive ball milling;* J. Alloys Compd **2007**, *427*, 204-208.

(165) L. Lutterotti, MAUD software, www.ing.unitn.it/~maud/, **2009**

(166) Brydson, R.; Sauer, H.; Engel, W.; Zeitler, E. *EELS as a fingerprint of the chemical co-ordination of light elements;* Microsc. Microanal. Microsctruct. **1991**, *2*, 159-169.

(167) Nakamoto, K. *Infrared and Raman spectra of inorganic and coordination compounds, 4th Edition.* Wiley: New York, **1986**.

(168) Fagerlund, G. *Determination of specific surface by the BET method;* Mater. Struct. **1973**, *6*, 239-245.

(169) *Encyclopedia of Chemical Physics and Physical Chemistry,* Vol. Volume II: Methods, IoP (Institute of Physics) Publishing: Bristol, UK, **2009**.

(170) Höhne, G.; Hemminger, W.; Flammershein, J. J. *Differential Scanning Calorimetry: an introduction for practitioners,* Springer: Berlin Heidelberg, **1996**.

(171) van Zeppelin, F.; Haluska, M.; Hirscher, M. *Thermal desorption spectroscopy as a quantitative tool to determine the hydrogen content in solids;* Thermochim. Acta **2003**, *404*, 251-258.

(172) Soni, P. R. *Mechanical alloying. Fundamentals and applications.,* Cambridge International Science Publishing: Cambridge, UK, **2000**.

(173) Wang, J.; Ebner, A. D.; Ritter, J. A. *Synthesis of metal complex hydrides for hydrogen storage;* J. Phys. Chem. C **2007**, *111*, 14917-14924.

(174) Magini, M.; Colella, C.; Iasonna, A.; Padella, F. *Power measurements during mechanical milling - II. The case of "single path cumulative" solid state reaction;* Acta Mater. **1998**, *46*, 2841-2850.

(175) Pitt, M. P.; Vullum, P. E.; Sorby, M. H.; Sulic, M. P.; Jensen, C. M.; Walmsley, J. C.; Holmestad, R.; Hauback, B. C. *Structural properties of the nanoscopic $Al_{85}Ti_{15}$ solid solution observed in the hydrogen-cycles $NaAlH_4$ + 0.1 $TiCl_3$ system;* Acta Mater. **2008**, *56*, 4691-4701.

(176) Sun, D.; Srinivasan, S. S.; Chen, G.; Jensen, C. M. *Rehydrogenation and cycling studies of dehydrogenated $NaAlH_4$;* J. Alloys Compd **2004**, *373*, 265-269.

(177) Chen, Y.; Williams, J. S. *Investigation of gas-solid reactions realised by ball milling;* Mater. Sci. Forum **1996**, 545-551.

(178) Campbell, C. T.; Peden, H. F. *Oxygen vacancies and catalysis on ceria surface;* Science **2005**, *309*, 713-714.

(179) Wang, P.; Kang, X. D.; Cheng, H. M. *Exploration of the nature of active Ti species in metallic Ti-doped $NaAlH_4$;* J. Phys. Chem. B **2005**, *109*, 20131-20136.

(180) Ocko, M.; Babic, E.; Zlatic, V. *Changes of the lattice parameter in Al 3d alloys due to virtual bound state;* Solid State Commun. **1976**, *18*, 705-708.

(181) Zaluska, A.; Zaluski, L.; Ström-Olsen, J. O. *Synergy of hydrogen sorption in ball-milled hydrides of Mg and Mg_2Ni;* J. Alloys Compd **1999**, *289*, 197-206.

(182) Baldé, C. P.; Hereijgers, B. P. C.; Bitter, J. H.; de Jong, K. P. *Facilitated hydrogen storage in $NaAlH_4$ supported on carbon nanofibers;* Angew. Chem. Int. Ed **2006**, *45*, 3501-3503.

(183) Schmiedgen, M.; Graat, P. C. J.; Baretzky, B.; Mittemeijer, E. J. *The initial stages of oxidation of gamma-TiAl: an X-ray photoelectron study;* Thin Solid Films **2002**, *415*, 114-122.

(184) Hamer, A. D.; Tisley, D. G.; Walton, R. A. *The x-ray photoelectron spectra of inorganic molecules - 8 : Compounds of scandium (III), scandium (III) oxide and complexes with organic oxygen donor molecules.;* J. Inorg. Nucl. Chem. **1974**, *36*, 1771-1775.

(185) Hwu, S. J.; Corbett, J. D.; Poeppelmeier, K. R. *Interstitial atoms in metal-metal bonded arrays: The synthesis and characterization of heptascandium decachlorodicarbide, $Sc_7Cl_{10}C_2$, and comparison with the interstitial-free Sc_7Cl_{10};* J. Solid State Chem. **1985**, *57*, 43-58.

(186) Praline, G.; Koel, B. E.; Hance, R. L.; Lee, H. I.; White, J. M. *X-ray photoelectron study of the reaction of oxygen with cerium;* J. Electron Spectrosc. Relat. Phenom. **1980**, *21*, 17-30.

(187) *CRC Handbook of Chemistry and Physics, 89th edition.* Boca Raton (USA), **2008**.

(188) Sandia National Laboratory, Hydride Properties Database, http://hydpark.ca.sandia.gov/PropertiesFrame.html, **2009**

(189) Hashi, K.; Ishikawa, K.; Suzuki, K.; Aoki, K. *Hydrogen absorption and desorption in the binary Ti-Al system;* J. Alloys Compd **2002**, *330-332*, 547-550.

(190) Spatz, P.; Gross, K. J.; Züttel, A.; Schlapbach, L. *Hydriding properties of Ce(Mn, Al)$_2$ and Ce(Fe, Al)$_2$ intermetallic compounds;* J. Alloys Compd **1997**, *260*, 211-216.

(191) *CRC Handbook of Chemistry and Physics, 75th edition.* **Boca Raton (USA), 1994**.

(192) Ma, L. P.; Kang, X. D.; Dai, H. B.; Liang, Y.; Fang, Z. Z.; Wang, P. J.; Wang, P.; Cheng, H. M. *Superior catalytic effect of TiF_3 over $TiCl_3$ in improving the hydrogen sorption kinetics of MgH_2: catalytic role of fluorine anion;* Acta Mater. **2009**, *57*, 2250-2258.

(193) Agresti, F.; Khandelwal, A. *Evidence of formation of $LiBH_4$ by high-energy ball milling of LiH and B in a hydrogen atmosphere;* Scr. Mater. **2009**, *60*, 753-755.

(194) Renaudin, G.; Gomes, S.; Hagemann, H.; Keller, L.; Yvon, K. *Structural and spectroscopic studies on the alkali borohydrides MBH_4 (M= Na, K, Rb, Cs);* J. Alloys Compd **2004**, *375*, 98-106.

(195) Borgschulte, A.; Züttel, A.; Hug, P.; Racu, A. M.; Schoenes, J. *Hydrogen-Deuterium exchange in bulk $LiBH_4$;* J. Phys. Chem. A **2008**, *112*, 4749-4753.

(196) Emery, A. R.; Taylor, R. C. *Raman spectroscopy in liquid ammonia solutions. Vibrational frequencies and force constants for isotopic species of the borohydride ion having tetrahedral symmetry;* J. Chem. Phys. **1958**, *28*, 1029-1032.

(197) Tse, J. S.; Klug, D. D.; Desgreniers, S.; Smith, J. S.; Flacau, R.; Liu, Z.; Hu, J.; Chen, N.; Jiang, D. T. *Structural phase transition in CaH_2 at high pressures;* Phys. Rev. B **2007**, *75*, 134108.

(198) Xu, T. T.; Zhen, J. G.; Nicholls, A. W.; Stankovich, S.; Piner, R. D.; Ruoff, R. S. *Single-crystal calcium hexaboride nanowires: synthesis and characterization;* Nano Lett. **2004**, *4*, 2051-2055.

(199) Fang, Z. Z.; Ma, L. P.; Kang, X. D.; Wang, P. J.; Wang, P.; Cheng, H. M. *In situ formation and rapid decomposition of $Ti(BH_4)_3$ by mechanical milling $LiBH_4$ with TiF_3;* Appl. Phys. Lett. **2009**, *94*, 044104.

(200) Pendolino, F.; Mauron, Ph.; Borgschulte, A.; Züttel, A. *Effect of boron on the activation energy of the decomposition of $LiBH_4$;* J. Phys. Chem. C **2009**, *113*, 17231-17234.

(201) Brinks, H. W.; Fossdal, A.; Hauback, B. C. *Adjustement of the stability of complex hydrides by anion substitution;* J. Phys. Chem. C **2008**, *112*, 5658-5661.

(202) Yin, L. C.; Wang, P.; Fang, Z. Z.; Cheng, H. M. *Thermodynamically tuning $LiBH_4$ by fluorine anion doping for hydrogen storage : a density functional study;* Chem. Phys. Lett. **2008**, *450*, 318-321.

(203) Wilson-Short, G. B.; Janotti, A.; Peles, A.; Van de Walle, C. G. *First-prinicples investigations of F and Cl impurities in* $NaAlH_4$; J. Alloys Compd **2009**, *484*, 347-351.

(204) Boesenberg, U.; Vainio, U.; Pranzas, P. K.; Bellosta von Colbe, J. M.; Goerigk, G.; Welter, E.; Dornheim, M.; Schreyer, A.; Bormann, R. *On the chemical state and distribution of Zr- and V-based additives in reactive hydride composites;* Nanotechnology **2009**, *20*, 204003.

(205) Dobrovolsky, V. D.; Ershova, O. G.; Solonin, Y. M.; Khyzhun, O. Y.; Paul-Boncour, V. *Influence of* TiB_2 *addition upon thermal stability and decomposition temperature of the* MgH_2 *hydride of a Mg-based mechanical alloy;* J. Alloys Compd **2008**, *465*, 177-182.

(206) Na Ranong, C.; Höhne, M.; Franzen, J.; Hapke, J.; Fieg, G.; Dornheim, M.; Eigen, N.; Bellosta von Colbe, J. M.; Metz, O. *Concept, design and manufacture of a prototype hydrogen storage tank based on sodium alanate;* Chem. Eng. Technol. **2009**, *32*, 1154-1163.

(207) Barkhordarian, G.; Klassen, T.; Dornheim, M.; Bormann, R. *Unexpected kinetic effect of* MgB_2 *in reactive hydride composites containing complex borohydrides;* J. Alloys Compd **2007**, *440*, L18-L21.

(208) Li, H. W.; Orimo, S.; Nakamori, Y.; Miwa, K.; Ohba, N.; Towata, S.; Züttel, A. *Materials designing of metal borohydrides: viewpoints from thermodynamical stabilities;* J. Alloys Compd **2007**, *446-47*, 315-318.

(209) Ravnsbaek, D.; Filinchuk, Y.; Cerenius, Y.; Jakobsen, H. J.; Besenbacher, F.; Skibsted, J.; Jensen, T. R. *A series of mixed-metal borohydrides;* Angew. Chem. **2009**, *48*, 6659-6663.

(210) Lindemann, I.; Domenech Ferrer, R.; Dunsch, L.; Filinchuk, Y.; Cerny, R.; Hagemann, H.; D'Anna, V.; Lawson Daku, L. M.; Schultz, L.; Gutfleisch, O. $Al_3Li_4(BH_4)_{13}$: *a complex double-cation borohydride with a new structure;* Chem. Eur. J. **2010**, *16*, 8707-8712.

(211) Cerny, R.; Severa, G.; Ravnsbaek, D.; Filinchuk, Y.; D'Anna, V.; Hagemann, H.; Haase, D.; Jensen, C. M.; Jensen, T. R. $NaSc(BH_4)_4$: *a novel scandium-based borohydride;* J. Phys. Chem. C **2010**, *114*, 1357-1364.

(212) Mosegaard, L.; Moller, B.; Jorgensen, J. E.; Filinchuk, Y.; Cerenius, Y.; Hanson, J. C.; Dimasi, E.; Besenbacher, F.; Jensen, T. R. *Reactivity of* $LiBH_4$: *in situ synchrotron radiation X-ray diffraction study;* J. Phys. Chem. C **2008**, *112*, 1299-1303.

(213) Arnbjerg, L. M.; Ravnsbaek, D.; Filinchuk, Y.; Vang, R. T.; Cerenius, Y.; Besenbacher, F.; Jorgensen, J. E.; Jakobsen, H. J.; Jensen, T. R. *Structure and dynamics for* $LiBH_4$-$LiCl$ *solid solutions;* Chem. Mater. **2009**, *21*, 5772-5782.

(214) Ketelaar, J. A. A.; Schutte, C. J. H. *The borohydride ion* $(BH_4)^-$ *in a face-centered cubic alkali-halide lattice;* Spectrochim. Acta **1961**, *17*, 1240-1243.

(215) Gao, J.; Adelhelm, P.; Verkuijlen, M. H. W.; Rongeat, C.; Herrich, M.; van Bentum, P. J. M.; Gutfleisch, O.; Kentgens, A. P. M.; de Jong, K. P.; de Jongh, P. E. *Confinement of $NaAlH_4$ in nanoporous carbon: Impact on H_2 release, reversibility and thermodynamics;* J. Phys. Chem. C **2010**, *114*, 4675-4682.

(216) Lohstroh, W.; Roth, A.; Hahn, H.; Fichtner, M. *Thermodynamic effects in nanoscale $NaAlH_4$;* Chem. Phys. Chem. **2010**, *11*, 789-792.

(217) Zhang, Y.; Zhang, W.; Wang, A.; Sun, L.; Fan, M.; Chu, H.; Sun, J.; Zhang, T. *$LiBH_4$ nanoparticles supported by disordered mesoporous carbon: hydrogen storage performances and destabilization mechanisms.;* Int. J. Hydrog. Energy **2007**, *32*, 3976-3980.

(218) Vajo, J. J.; Olson, G. L. *Hydrogen storage in destabilized chemical systems;* Scr. Mater. **2007**, *56*, 829-834.

(219) Ingleson, M. J.; Perrez Barrio, J.; Bacsa, J.; Steiner, A.; Darling, G. R.; Jones, J. T. A.; Khimyak, Y. Z.; Rosseinsky, M. J. *Magnesium borohydride confined in a metal-organic framework: a preorganized system for facile arene hydroboration;* Angew. Chem. Int. Ed **2009**, *48*, 2012-2016.

(220) Fichtner, M.; Zhao-Karger, Z.; Hu, J.; Roth, A.; Weidler, P. *The kinetic properties of $Mg(BH_4)_2$ infiltrated in activated carbon;* Nanotechnology **2009**, *20*, 204029.

(221) Adelhelm, P.; Gao, J.; Verkuijlen, M. H. W.; Rongeat, C.; Herrich, M.; van Bentum, P. J. M.; Gutfleisch, O.; Kentgens, A. P. M.; de Jong, K. P.; de Jongh, P. E. *Comprehensive study of melt infiltration for the synthesis of $NaAlH_4/C$ nanocomposites;* Chem. Mater. **2010**, *22*, 2233-2238.

(222) Lee, Y. S.; Kim, Y.; Cho, Y. W.; Shapiro, D.; Wolverton, C.; Ozolins, V. *Crystal structure and phonon instability of high-temperature β-$Ca(BH_4)_2$;* Phys. Rev. B **2009**, *79*, 104107.

7. Annexes

A. Crystal structures of $NaAlH_4$ and Na_3AlH_6

Phase	Space group	Cell parameters (Å)	Properties	Scheme	References
$NaAlH_4$	$I4_1/a$ (tetragonal)	a=5.021 c=11.346	• Melts at about 180°C. • Decomposes in Na_3AlH_6 from melt above 207°C when undoped	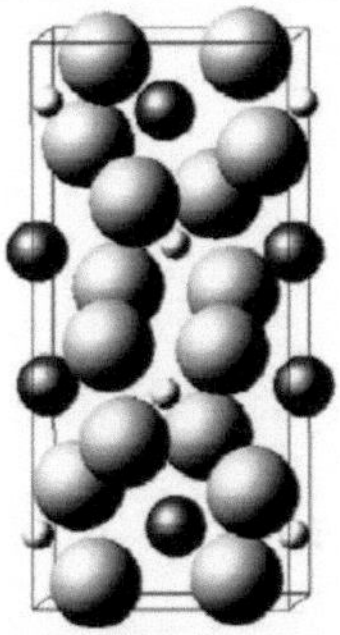	[49] [46] [47]
α-Na_3AlH_6	$P2_1/n$ (monoclinic)	a=5.408 b=5.538 c=7.757 β=89.83°	Transforms in β phase (endothermic reaction) at 252°C	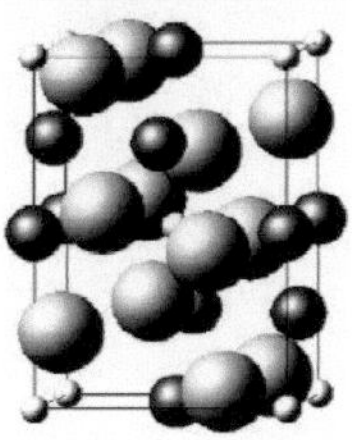	[49] [47] [48]
β-Na_3AlH_6	Fm-3m (cubic)	a=7.755	Decomposes in NaH above 270°C when undoped	-	[47] [49] [48]

B. Crystal structures of $Ca(BH_4)_2$ polymorphs and intermediate phases

Phase	Space group	Cell parameters (Å)	Properties	Scheme	References
α	Fddd (orthorhombic)	a=8.791 b=13.137 c=7.500	Transforms in β phase (endothermic reaction) at 140-170°C (reversible)	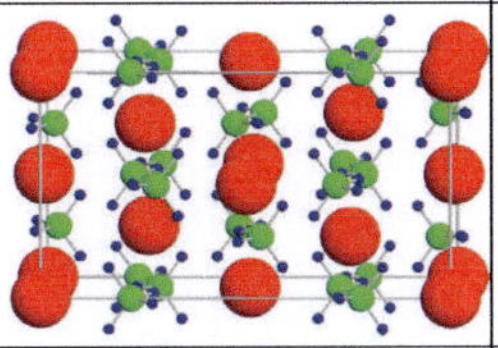	[31] [145] [222] [155] [156]
	F2dd (orthorhombic)	a=8.7759 b=13.0234 c=7.4132	• Transforms to α' above 220°C • Starts gradually transformation to β phase at 180°C	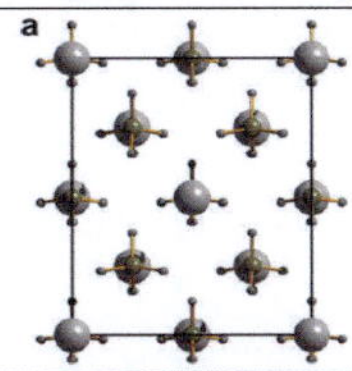	[151] [152]
	C2/c (monoclinic)	a=7.51 b=8.70 c=7.50 β=119.35°		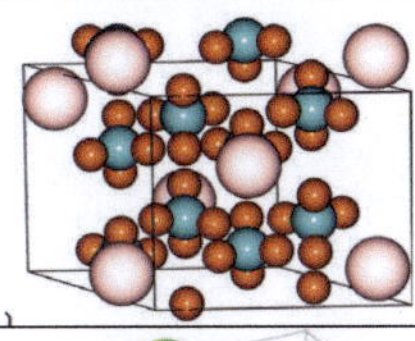	[149]
β	$P4_2/m$ (tetragonal)	a=6.9468 c=4.3661	• High temperature phase but can be observed at room temperature • Decomposes in two steps at 350-380°C and 400-500°C	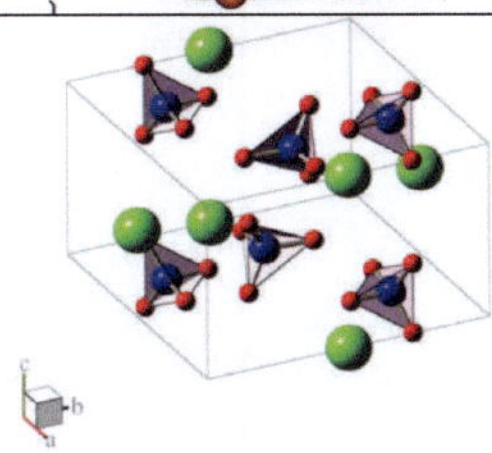	[148] [222] [152] [156] [146] [145]
	P-4 (tetragonal)	a=6.91894 c=4.34711	Decomposes at *ca.* 380°C	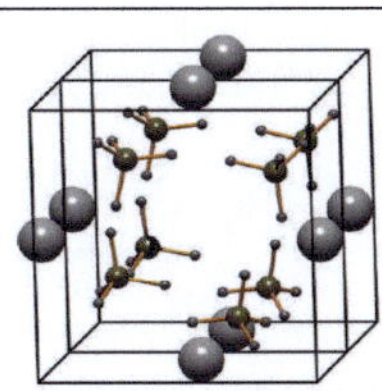	[151] [149]

γ	Pbca (orthorhombic)	a=13.0584 b=8.3881 c=7.5107	• Present at room temperature • Transforms to β phase above 170°C (not reversible)	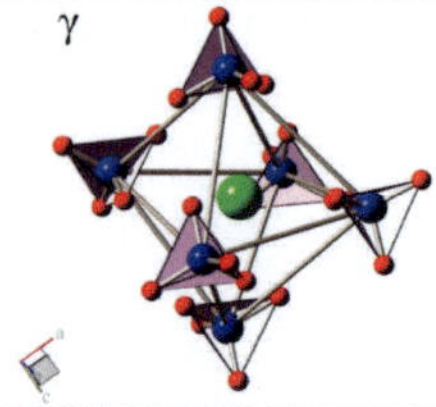	[148] [150] [149] [145]
δ	Not determined	Not determined	From the transformation of γ phase at 400-480°C	-	[146]
α'	I-42d (tetragonal))	a=5.8446 b=13.2279	From second order transformation of α at 222°C	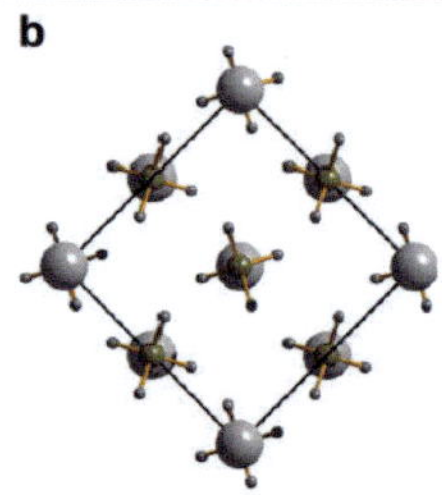	[151]
$Ca(BH_4)_2.H_2O$	Pnma (orthorhombic)	a=8.200 b=5.8366 c=11.851	From water contamination	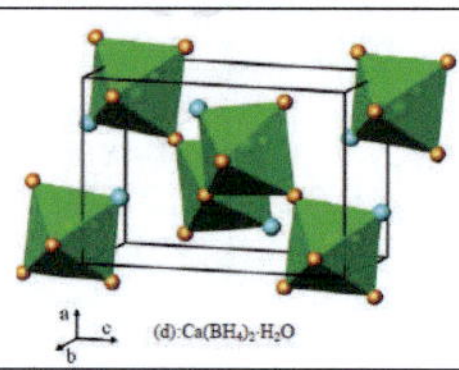	[145]
CaB_2H_2	Pnma (orthorhombic)	a=12.8105 b=4.0825 c=3.9034	• Intermediate phase. • Exits between 370 and 400°C	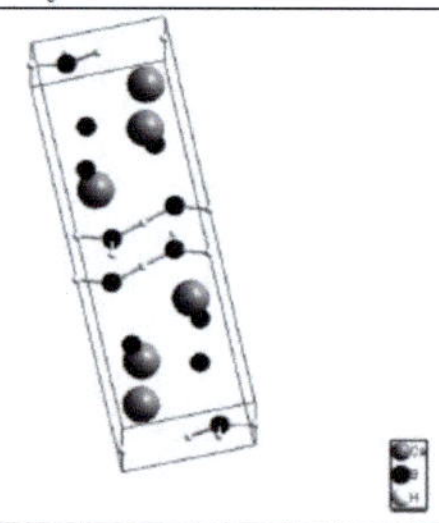	[160]

C. Summary and organisation of the results presented in this manuscript

C1: Organisation of Chapter 3

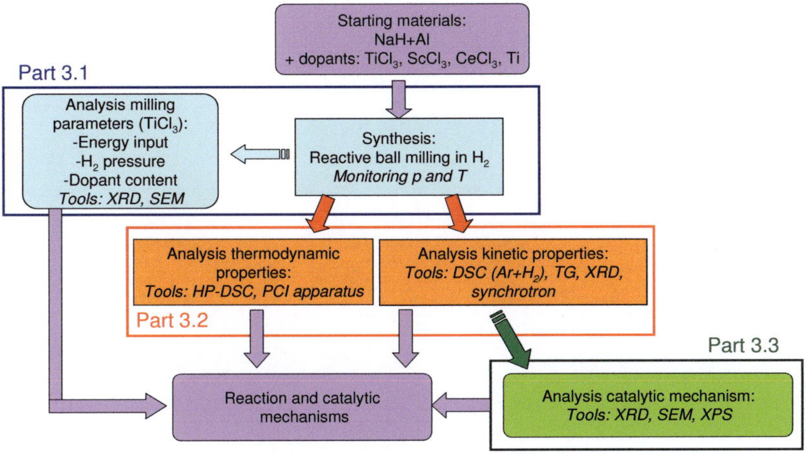

C2: Organisation of Chapter 4

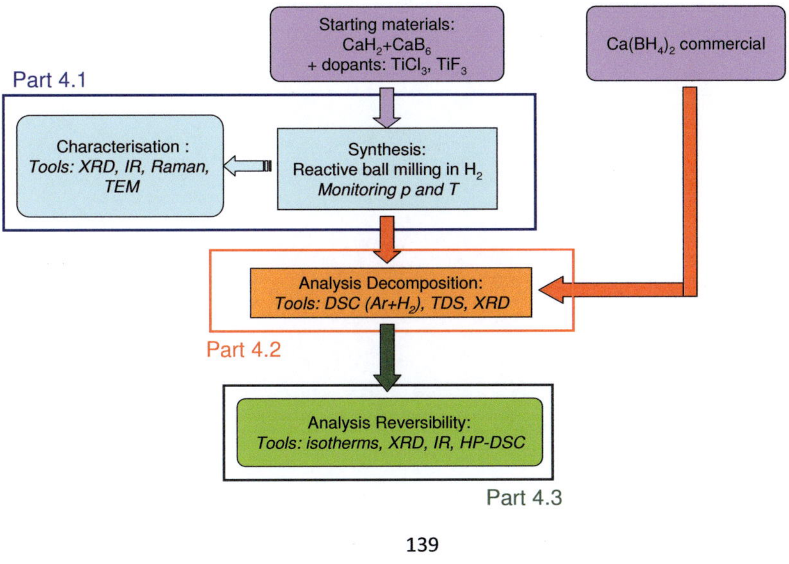

D. Summary of the properties obtained for doped-$NaAlH_4$

Dopant		No[a]	$TiCl_3$			$ScCl_3$	$CeCl_3$	Ti
Content (mol%)		**0**	**2**	**4**	**6**	**4**	**4**	**4**
Reaction yield after 5 h milling[b,c] (%)		-	70	94	99	84	89	26
Max. H_2 capacity (Theo.)[b,c] (wt%)		5.6	5.30	5.03	4.78	5.03	4.73	5.41
Max. H_2 capacity (Theo.)[b,d] (wt%)		-	5.00	4.45	3.96	4.46	4.19	-
Plateau pressure in abs-/desorption for reaction (1.6) (bar H_2)		-	-	6.5/2.3	-	8.3/2.3	6.5/1.1	-
Plateau pressure in abs-/desorption for reaction (1.7) (bar H_2)		-	-	83.1/57.6	-	80.2/57.6	82.0/57.6	-
Max. H_2 capacity measured[b] (wt%) TG (*yield[b,d]*)		4.63 (*83%*)	3.34 (*67%*)	3.64 (*82%*)	2.21 (*56%*)	3.74 (*84%*)	2.65 (*63%*)	1.36 (*26%*)
Decomposition temp. (DSC Ar) (°C)		242	115	110	101	125	125	-
Re-absorption temp. (DSC 50 bar H_2) (°C)		-	-	154	-	87	62	-
Surface area BET (m^2 g^{-1}): as-milled/heated to 300°C		-	-	8.3/6.6	-	-	9.6/2.8	-
Chemical state dopant (XPS) for as-milled sample (0/17 min sputtering)		-	-	TiO/Ti	-	$ScCl_3$+ Sc_2O_3/ $ScCl_3$+Sc	Ce^{III}/ Ce^{III}+Ce	TiO/Ti
Surface comp.[e] (%at) for as-milled sample (0/17 min sputtering)	O	45.6/27.4	-	39.7/25.4	-	40.9/20.8	40.5/29.8	47.1/41.5
	Na	31.8/31.9		40.3/31.1		31.7/28.5	35.7/30.5	42.5/39.3
	Al	22.6/40.7		15.2/37.7		23.1/43.9	20.1/33.2	10/24.3
	M[f]	-		0.5/1.3		0.8/2	0.9/1.5	0.4/0.4
	Cl	-		4.3/4.6		3.6/4.7	2.8/4.4	-
Chemical state dopant (XPS) after heating to 300°C (0/17 min sputtering)		-	-	Ti/Ti	-	$ScCl_3$+ Sc_2O_3/ $ScCl_3$+Sc	Ce^{III}/ Ce^{III}+Ce	-
Surface comp.[e] (%at) after heating to 300°C (0/17 min sputtering)	O	-	-	37.8/34.7	-	38.4/31.1	38.6/31.7	-
	Na			40.1/29.9		39.9/28.6	37.7/32.7	
	Al			18.8/29.6		18.5/36	20.9/32	
	M[f]			0.4/0.9		0.7/1.3	0.5/1.5	
	Cl			2.8/4.8		2.5/3	2.2/2.8	

[a] $NaAlH_4$ from Aldrich (90% purity)
[b] for reactions (1.6) and (1.7)
[c] considering only weight of dopant added
[d] assuming formation of elemental Ti, Sc or Ce and NaCl from the decomposition of the dopants
[e] surface composition from XPS analysis, ca. 10% precision
[f] M=Ti, Sc or Ce

E. Indexation electron diffraction patterns (TEM)

E1) Figure 4.7d

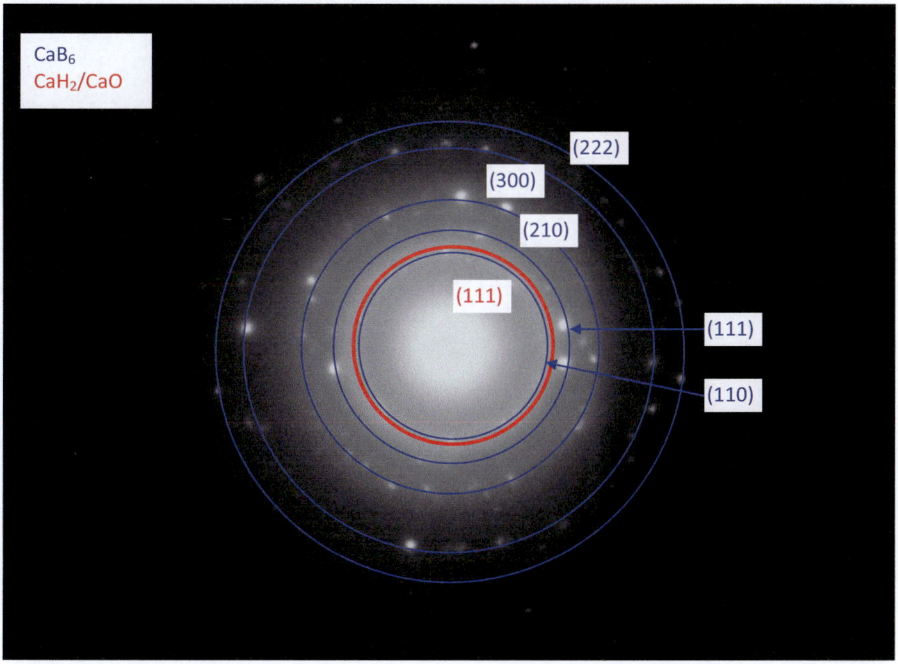

E2)Figure 4.8b

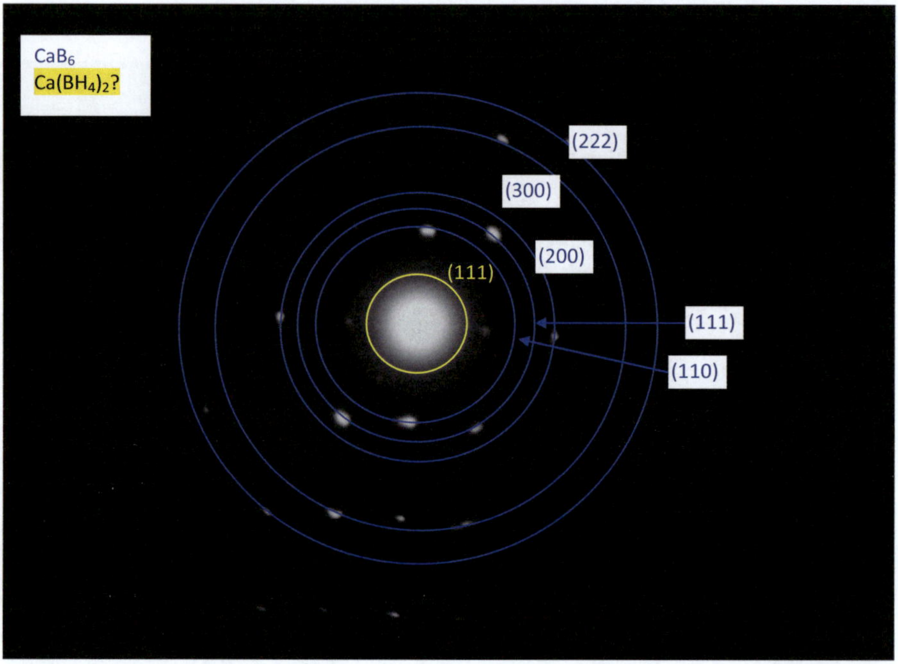

List of publications

As a first author:

C. Rongeat, I. Llamas-Jansa, S. Doppiu, S. Deledda, A. Borgschulte, L. Schultz and O. Gutfleisch – *Determination of the heat of hydride formation/decomposition by high-pressure differential scanning calorimetry (HP-DSC)* – Journal of Physical Chemistry B – **111** (2007) 13301

C. Rongeat, C. Geipel, I. Llamas-Jansa, L. Schultz and O. Gutfleisch – *Influence of the dopant during the one-step mechano-chemical synthesis of sodium alanate* – Journal of Physics: Conferences Series – **144** (2009) 012022

C. Rongeat, I. Llamas Jansa, S. Oswald, L. Schultz and O. Gutfleisch – *Mechanochemical synthesis and XPS analysis of sodium alanate with different additives* – Acta Materialia - **57** (2009) 5563

C. Rongeat, V. D'Anna, H. Hagemann, A. Borgschulte, A. Züttel, L. Schultz and O. Gutfleisch – *Effect of additives on the synthesis and reversibility of* $Ca(BH_4)_2$ – Journal of Alloys and Compounds - **493** (2010) 281

C. Rongeat, I. Lindemann, A. Borgschulte, L. Schultz and O. Gutfleisch – *Effect of the presence of halides on the synthesis and decomposition of* $Ca(BH_4)_2$ – Internation Journal of Hydrogen Energy – **36** (2011) 247

C. Rongeat, N. Scheerbaum, L. Schultz and O. Gutfleisch – *Catalysis of* H_2 *sorption in* $NaAlH_4$*: General description and new insights* – Acta Materialia – **59** (2011) 1725

As a co-author:

I. Llamas-Jansa, **C. Rongeat**, S. Doppiu, O. Gutfleisch and L. Schultz – *Characterisation of complex hydrides synthesised or modified by ball milling* – International Journal of Materials Research - **99** (2008) 553

I. Llamas-Jansa, **C. Rongeat**, S. Doppiu, O. Gutfleisch - *synthesis and modification of light metal and complex hydrides by high-energy ball milling* - in Materials issues in a hydrogen economy. Proceedings of the International Symposium, Richmond (USA), Nov. 12-15, 2007 - pp. 138-143 - World Scientific Publishing, Singapore, 2009.

P. Adelhelm, J. Gao, M. Verkuijlen, **C. Rongeat**, M. Herrich, P. van Bentum, O. Gutfleisch, A. Kentgens, K. de Jong, P. de Jongh - *Comprehensive study of melt infiltration for the synthesis of NaAlH4/C nanocomposites* - Chemistry of Materials – **22** (2010) 2233

J. Gao, M. Verkuijlen, **C. Rongeat**, M. Herrich, O. Gutfleisch, K. de Jong, P. de Jongh, P. Adelhelm, J. van Betum, A. Kentgens – *Confinement of $NaAlH_4$ in nanoporous carbon: impact on H_2 release, reversibility and thermodynamics* –Journal of Physical Chemistry C – **114** (2010) 4675

J. Paillier, **C. Rongeat**, O. Gutfleisch, A. Gebert - *Hydrogen and Zr-based Metallic Glasses: Gas/Solid Absorption Process and Structure Evolution* – Journal of Alloys and Compounds – 509 (2011) 1636

C. Bonatto Minella, S. Garoni, C. Pistidda, R. Goslawit-Utke, G. Barkhordarian, **C. Rongeat**, I. Lindemann, O. Gutfleisch, T.R. Jensen, Y. Cerenius, J. Christensen, M.D. Baro, R. Bormann, T. Klassen, M. Dornheim – *Effect of transition metal fluorides on the sorption properties and reversible formation of $Ca(BH_4)_2$* – Journal of Physical Chemistry C – (2011) in press, doi:10.1021/jp107781m

List of figures

Chapter 1:

Figure 1.1: Hydrogen cycle 2

Figure 1.2: Principle of a PEM fuel cells 3

Figure 1.3: State-of-the-art of hydrogen storage technologies 6

Figure 1.4: Phase diagram of a conventional metal hydride 9

Figure 1.5: PCI curves for NaAlH4 doped with $Ti(OBu)_4$ 12

Figure 1.6: DSC traces for undoped $NaAlH_4$ and for doped $NaAlH_4$ 14

Figure 1.7: Dehydrogenation curves of doped-$NaAlH_4$ with different chloride dopants 16

Figure 1.8: Sketch of $NaAlH_4$ decomposition via vacancies diffusion 20

Figure 1.9: van't Hoff plots for different reactions 22

Figure 1.10: Different $Ca(BH_4)_2$ polymorphs 24

Figure 1.11: DSC traces obtained during heating of $Ca(BH_4)_2$. 25

Chapter 2:

Figure 2.1: Scheme of the milling vial used for reactive ball milling 30

Figure 2.2: SEM pictures for as-milled sample NaAlH4 with 4mol% TiCl3 33

Figure 2.3: Normal vibrational modes in XY_4 tetrahedral molecule 35

Figure 2.4: Adsorption isotherm for N_2 at 77K and BET plot 37

Figure 2.5: DSC traces for $TiCl_3$ doped-Na_3AlH_6 at 120 bar H_2 39

Figure 2.6: Example of PCI curves and van't Hoff plot for $LaNi_5$ 42

Chapter 3:

Figure 3.1: Variation of pressure and temperature during milling 47

Figure 3.2: XRD patterns for the starting powder mixture, and as-milled powders milled. 48

Figure 3.3: SEM pictures of as-milled $NaAlH_4$ with $TiCl_3$ as dopant 49

Figure 3.4: Variations of the pressure and the temperature for different starting pressures 51

Figure 3.5: XRD patterns for the samples milled with different H_2 starting pressures 52

Figure 3.6: Variations of the pressure and the temperature with different quantity of $TiCl_3$ 53

Figure 3.7: XRD patterns for as-milled samples with different dopant content 54

Figure 3.8: Variations of the pressure and the temperature for different dopants 55

Figure 3.9: XRD patterns of the samples milled with different dopants 56

Figure 3.10: In-situ synchrotron measurements during the first decomposition 59

Figure 3.11: DSC traces in Ar for undoped $NaAlH_4$ and $NaAlH_4$ + 4mol% $CeCl_3$ 60

Figure 3.12: DSC traces for doped-NaAlH4 for increasing H_2 pressures 62

Figure 3.13: van't Hoff plots obtained from isobaric DSC 64

Figure 3.14: PCI curves measured at 150°C for $NaAlH_4$ 65
Figure 3.15: van't Hoff plots obtained from PCI curves 66
Figure 3.16: DSC traces for $NaAlH_4$ with different dopants measured in Ar. 67
Figure 3.17: DSC traces and TG curves for doped $NaAlH_4$ 68
Figure 3.18: XRD patterns measured after different storage time 69
Figure 3.19: DSC traces during cycling at 3 and 50 bar H_2 70
Figure 3.20: In-situ synchrotron measurements for sample $NaAlH_4$ with 4mol% $CeCl_3$ 71
Figure 3.21: XRD patterns for the decomposed samples 74
Figure 3.22: SEM pictures of sample $NaAlH_4$ doped with 4mol% $TiCl_3$ 75
Figure 3.23: SEM pictures for $NaAlH_4$ with 4mol% $CeCl_3$ 76
Figure 3.24: XPS spectra for Ti 2p, Sc 2p and Ce 3d for as-milled samples 78
Figure 3.25: XPS spectra for Ti 2p, Sc 2p and Ce 3d after heating in-situ to 300°C 79

Chapter 4:

Figure 4.1: Variations of the pressure and temperature for the first milling attempt 85
Figure 4.2: Variation of the quantity of H_2 and of temperature for improved milling conditions 86
Figure 4.3: XRD patterns for as-milled mixture 2 CaH_2 + CaB_6 + 0.08 additive 87
Figure 4.4: in-situ XRD measurements for as-milled samples with $TiCl_3$ and TiF_3 88
Figure 4.5: Raman spectra of as-received compounds and for the as-milled samples 89
Figure 4.6: IR spectra for $Ca(BH_4)_2$, CaH_2, CaB_6 and as-milled samples 90
Figure 4.7: TEM analysis (low magnification) of as-milled sample milled with TiF_3 92
Figure 4.8: TEM micrograph (high magnification) and Ca L edge and B K edge EELS spectra 93
Figure 4.9: DSC traces for as-received $Ca(BH_4)_2$ in different atmosphere 95
Figure 4.10: DSC traces in 3 bar H_2 for pure $Ca(BH_4)_2$ and as-milled samples 96
Figure 4.11: TDS spectra of H_2 and B_2H_6 for as-received compounds and as-milled samples 98
Figure 4.12: XRD patterns for 2 CaH_2 + CaB_6 + 0.08 TiF_3 after ex-situ heating 100
Figure 4.13: First desorption and following re-hydrogenation at 100 bar H_2 and 350°C 102
Figure 4.14: DSC traces during cycling at 3 and 100 bar H_2 103
Figure 4.15: XRD patterns for re-hydrogenated samples 104
Figure 4.16: IR spectra of $Ca(BH_4)_2$, CaH_2, and of the re-hydrogenated 105
Figure 4.17: DSC traces in H_2 for $Ca(BH_4)_2$, as-milled and re-hydrogenated sample with TiF_3 106

List of Tables

Chapter 1:

Table 1.1: Classification and properties of different hydrogen storage technologies 4
Table 1.2: DOE targets (revised) for on-board hydrogen storage systems (Feb. 09) 5
Table 1.3: Enthalpies of reaction for equation (1. 6) and (1. 7) reported in literature. 12

Chapter 2:

Table 2.1: Milling parameters for the synthesis of $NaAlH_4$ from NaH + Al with dopant 30
Table 2 2: Milling parameters for the synthesis of $Ca(BH_4)_2$ from CaH_2 + CaB_6 with additives 31

Chapter 3:

Table 3.1: Pressure difference Δp between the initial and the final H2 pressure after milling 48
Table 3.2: Summary of the results obtained from Rietveld refinement for as-milled samples 53

Chapter 4:

Table 4.1: Summary of the variations in H2 quantity (Δn) after milling 81

Acknowledgements

First of all, I would like to thank Prof. Ludwig Schultz and Prof. Asuncion Fernandez Camacho who accepted to review this work.

I would like to thank in particular Prof. Schultz who gave me the opportunity to work in IFW Dresden.

I had the chance to participate to the Marie Curie program of the European community that offered me the possibility to work in two different laboratories in IFW and Empa (Switzerland) during my PhD time.

I am very grateful to Dr. Oliver Gutfleisch who has welcomed me in his group in IFW and who mainly supervised this work and to Prof. Andreas Züttel and Dr. Andreas Borgschulte with whom I spent a fruitful year in Empa during the exchange with IFW. Working in these two different institutions with inspiring people and nice working atmosphere was a wonderful experience.

I also gratefully acknowledge the European community through the Marie Curie Research training network COSY for the financial support and for giving me the opportunity to work in an international environment with a large support to improve my knowledge not only in science but also to build a great personal experience.

During my PhD work, I had the chance to collaborate with different groups who gave me the opportunity to use their facilities to improve my research work. I especially want to thank:

- Dr. Steffen Oswald in IFW for giving me the opportunity to obtain very nice results with XPS experiments,
- Dr. T.R. Jensen and his group in University of Aarhus (Denmark) for the synchrotron measurements,
- Emilie Deprez and Prof. Asuncion Fernandez Camacho in ICMSE (Spain) for the nice work with TEM,
- Vincenza D'Anna and Dr. Hans Hagemann in University of Geneva (Switzerland) for the very interesting and useful infrared spectroscopy measurements.

And also in IFW and Empa, I received a very useful experimental help from colleagues that I would like to thank very much:

- Monika Herrich, Bernhard Gebel, Christian Geipel, Inge Lindemann, Nils Scheerbaum in IFW,
- Andreas Borgschulte, Arndt Remhof, Oliver Friedrichs, Shunsuke Kato, Flavio Pendolino in Empa.

Having worked in two institutions, there is a long list of colleagues I would like to thank for their help with science or german/swiss, for nice chats and coffee breaks:

In IFW: Nils (my eldest officemate, alias Dr. Nils, thanks also for german support), Christian and Christian, Inge, Monika (thanks for the science and more…), Bernhard, Isabel, Stefania, Jule, Konrad, Jian, Maria, Tom, Julia and the other former and new members of the group.

In Empa: Flavio, Shunsuke, Corinne (swiss-english translation), Daniel, Michal, Andreas, Oliver, Arndt, Philippe, Robin, Michael, Pascal, Filippo, Riccarda, Zbigniew, Valentina, Ulrich, Jelena, Florian.

The nice thing with the European project is that you met nice people from all over Europe. I would like to thank all the COSY partners that I was really pleased to meet every six months for the very nice getting together and also the help with exchanging experience and problems.

In particular, I would like to wish all the best to my "COSY fellows" with whom I spent very nice moments and appreciate the support during this work:

Emilie (French connection, thanks for all), Christian (I was happy to have you in Dresden also), Sebastiano (sorry for the bad weather when you came to Dresden), Claudio (except the snowball), Shunsuke (Japanese food and salsa teacher) Francesco (thanks for the visit in Torino), Marta, Daphiny, Christopher, Deniz, Maria.

I want also to thank all the people I met during these last years that help me or for the very useful discussions in science and for the nice collaborations: Aline, Petra, Philipp, José, Ulrike, Gustavo, Kasia, Jörg, Flaviu, Jakub. I hope I forget nobody.

Last but not least, I would like to thank my family for their support all along this period of my life.

I like to thank in particular Jerome because he had the most difficult task: supporting me every days especially the unsuccessful ones, reading my papers and waste a very long time correcting this manuscript in a topic that can be really tedious.

Erklärung gemäß §6 Punkt 5a) bis d) der Promotionsordnung der Fakultät Maschinenwesen

Hiermit versichere ich, dass ich die vorliegende Arbeit ohne unzulässige Hilfe Dritter und ohne Benutzung anderer als der angegebenen Hilfsmittel angefertigt habe; die aus fremden Quellen direkt oder indirekt übernommenen Gedanken sind als solche kenntlich gemacht. Bei der Erstellung des Manuskripts habe ich Unterstützung von folgenden Personen erhalten:

Prof. Dr. Ludwig Schultz

Dr. Oliver Gutfleisch.

Weitere Personen waren an der geistigen Herstellung der vorliegenden Arbeit nicht beteiligt. Insbesondere habe ich nicht die Hilfe eines kommerziellen Promotionsberaters in Anspruch genommen. Dritte haben von mir keine geldwerten Leistungen für Arbeiten erhalten, die in Zusammenhang mit dem Inhalt der vorgelegten Dissertation stehen. Die Arbeit wurde bisher weder im Inland noch im Ausland in gleicher oder ähnlicher Form einer anderen Prüfungsbehörde vorgelegt und ist auch noch nicht veröffentlicht worden. Die Promotionsordnung der Fakultät Maschinenwesen der TU Dresden wird anerkannt.

Carine Rongeat   Dresden, den 19. August 2010